*History of Analytic Philosophy*

Series Editor: **Michael Beaney**

*Titles include:*

Omar W. Nasim
BERTRAND RUSSELL AND THE EDWARDIAN PHILOSOPHERS
Constructing the World

Pierre Wagner (*editor*)
CARNAP'S *LOGICAL SYNTAX OF LANGUAGE*

*Forthcoming:*

Andrew Arana and Carlos Alvarez (*editors*)
ANALYTIC PHILOSOPHY AND THE FOUNDATIONS OF MATHEMATICS

Sébastien Gandon
RUSSELL'S UNKNOWN LOGICISM
A Study in the History and Philosophy of Mathematics

Sandra Lapointe
BERNARD BOLZANO'S THEORETICAL PHILOSOPHY
An Introduction

Douglas Patterson
ALFRED TARSKI
Philosophy of Language and Logic

Erich Reck (*editor*)
THE HISTORIC TURN IN ANALYTIC PHILOSOPHY

Graham Stevens
THE THEORY OF DESCRIPTIONS

---

**History of Analytic Philosophy**
**Series Standing Order ISBN 978–0–230–55409–2 (hardcover)**
**Series Standing Order ISBN 978–0–230–55410–8 (paperback)**
(*outside North America only*)

You can receive future titles in this series as they are published by placing a standing order. Please contact your bookseller or, in case of difficulty, write to us at the address below with your name and address, the title of the series and the ISBN quoted above.

Customer Services Department, Macmillan Distribution Ltd, Houndmills, Basingstoke, Hampshire RG21 6XS, England

---

*Also by Pierre Wagner*

MATHÉMATIQUES ET EXPÉRIENCE, 1918–1940 (2008)
(Co-editor with Jacques Bouveresse)

LA LOGIQUE (2007)

L'ÂGE D'OR DE L'EMPIRISME LOGIQUE (2006)
(Co-editor with Christian Bonnet)

PHILOSOPHIE DES SCIENCES (2 vols, 2004)
(Co-editor with Sandra Laugier)

LES PHILOSOPHES ET LA SCIENCE (2002, editor)

LA MACHINE EN LOGIQUE (1998)

# Carnap's *Logical Syntax of Language*

Edited by

Pierre Wagner
*Université Paris 1 Panthéon-Sorbonne*
*IHPST, Paris*

palgrave
macmillan

First published 2009 by
PALGRAVE MACMILLAN

Palgrave Macmillan in the UK is an imprint of Macmillan Publishers Limited, registered in England, company number 785998, of Houndmills, Basingstoke, Hampshire RG21 6XS.

Palgrave Macmillan in the US is a division of St Martin's Press LLC, 175 Fifth Avenue, New York, NY 10010.

Palgrave Macmillan is the global academic imprint of the above companies and has companies and representatives throughout the world.

Palgrave® and Macmillan® are registered trademarks in the United States, the United Kingdom, Europe and other countries.

ISBN-13: 978–0–230–20151–4 hardback
ISBN-10: 0–230–20151–2 hardback

This book is printed on paper suitable for recycling and made from fully managed and sustained forest sources. Logging, pulping and manufacturing processes are expected to conform to the environmental regulations of the country of origin.

A catalogue record for this book is available from the British Library.

Library of Congress Cataloging-in-Publication Data
Carnap's Logical syntax of language / edited by Pierre Wagner.
p. cm. — (History of analytic philosophy)
Proceedings of a conference held Oct. 5–6, 2005 in Paris, France.
Includes bibliographical references (p. ) and index.
ISBN 978–0–230–20151–4
1. Carnap, Rudolf, 1891–1970. Logische Syntax der Sprache—Congresses. 2. Logic, Symbolic and mathematical—Congresses. 3. Grammar, Comparative and general—Syntax—Congresses. 4. Logical positivism—Congresses. I. Wagner, Pierre.
B945.C163L6338 2009
193—dc22 2008051553

10 9 8 7 6 5 4 3 2 1
18 17 16 15 14 13 12 11 10 09

Printed and bound in Great Britain by
CPI Antony Rowe, Chippenham and Eastbourne

# Contents

# Series Editor's Foreword

During the first half of the twentieth century analytic philosophy gradually established itself as the dominant tradition in the English-speaking world, and over the last few decades it has taken firm root in many other parts of the world. There has been increasing debate over just what 'analytic philosophy' means, as the movement has ramified into the complex tradition that we know today, but the influence of the concerns, ideas, and methods of early analytic philosophy on contemporary thought is indisputable. All this has led to greater self-consciousness among analytic philosophers about the nature and origins of their tradition, and scholarly interest in its historical development and philosophical foundations has blossomed in recent years, with the result that the history of analytic philosophy is now recognized as a major field of philosophy in its own right.

The main aim of the series in which the present book appears, the first series of its kind, is to create a venue for work on the history of analytic philosophy, consolidating the area as a major field of philosophy and promoting further research and debate. The 'history of analytic philosophy' is to be understood broadly, as covering the period from the last three decades of the nineteenth century to the start of the twenty-first century, beginning with the work of Frege, Russell, Moore, and Wittgenstein, who are generally regarded as its main founders, and the influences upon them, and going right up to the most recent developments. In allowing the 'history' to extend to the present, the aim is to encourage engagement with contemporary debates in philosophy, for example, in showing how the concerns of early analytic philosophy relate to current concerns. In focusing on analytic philosophy, the aim is not to exclude comparisons with other – earlier or contemporary – traditions, or consideration of figures or themes that some might regard as marginal to the analytic tradition but which also throw light on analytic philosophy. Indeed, a further aim of the series is to deepen our understanding of the broader context in which analytic philosophy developed, by looking, for example, at the roots of analytic philosophy in neo-Kantianism or British idealism, or the connections between analytic philosophy and phenomenology, or discussing the work of philosophers who were important in the development of analytic philosophy but who are now often forgotten.

The present book, edited by Pierre Wagner, is a collection of essays on one of the most important texts in the development of analytic philosophy, Rudolf Carnap's *Logische Syntax der Sprache* (*The Logical Syntax of Language*), first published in German in 1934 and in English in 1937. Influenced by Frege, Russell, and the early Wittgenstein, Carnap was a central figure in what can be regarded as the second generation of analytic philosophers, whose work came to maturity in the 1930s – the period in which analytic philosophy established itself as a movement. His first major work was *Der logische Aufbau der Welt* (*The Logical Structure of the World*), published in 1928, in which he sought 'rational reconstructions' of our empirical concepts based on what he called 'elementary experiences'. As the motto for his book he had taken Russell's famous 'maxim in scientific philosophising': 'Wherever possible, logical constructions are to be substituted for inferred entities'.

Carnap's project in the *Aufbau*, however, while not foundationalist in the traditional sense, was still conceived epistemologically. His *Logical Syntax*, on the other hand, represented a radical break with the past. Inspired by Wittgenstein's *Tractatus*, it marked Carnap's own 'linguistic turn', and influenced as well by Hilbert's idea of metamathematics and the developments in logic in the 1930s, in which Carnap himself played a leading role, it was Carnap's first attempt to respond to Wittgenstein's strictures on what could meaningfully be said by distinguishing between the 'material mode' and 'formal mode' of speech. Rejecting metaphysics, what was left of philosophy was identified with the 'logic of science', in turn understood as 'the logical syntax of the language of science'. A sentence in the material mode such as '5 is a number', for example, which might generate metaphysical questions as to the nature of numbers, was to be translated into the formal mode as ' "5" is a numeral', whose meaning was to be clarified by elucidating the role it plays in an arithmetical language. Carnap allowed, however, that there might be various possible languages, the decision between them being a pragmatic one based on their utility in science. The *Logical Syntax* contains the first articulation of Carnap's famous 'principle of tolerance':

> *In logic, there are no morals*. Everyone is at liberty to build up his own logic, i.e. his own form of language, as he wishes. All that is required of him is that, if he wishes to discuss it, he must state his methods clearly, and give syntactical rules instead of philosophical arguments. (*Logical Syntax*, p. 52)

Shortly after the publication of the *Logical Syntax*, influenced by Tarski's work on truth, Carnap underwent a 'semantic turn' and rejected the exclusive emphasis on the syntactic approach. But he retained his belief in the principle of tolerance and his general conception of philosophy, which became refined further in distinguishing between internal and external questions and in his discussions of 'explication'. So the *Logical Syntax* can indeed be regarded as the first flowering of Carnap's mature philosophy, and is important, too, for the influence it had on many subsequent philosophers, from his Vienna Circle colleagues, as well as Ayer and Quine, onwards. While Carnap's *Aufbau* has received a lot of attention over the last decade or so, and

Carnap's semantic works (such as *Meaning and Necessity*) have never dropped from the radar (even if they have only remained, regrettably, as the target of later critics such as Quine), the *Logical Syntax* has been relatively neglected. The present volume, the first book devoted to the work, puts the *Logical Syntax* firmly back in the pantheon of analytic philosophy. With a substantial introduction by the editor, clearly and helpfully explaining the context and content of the work, and rich and insightful contributions from many of the leading scholars of Carnap's philosophy, covering all aspects of the work, this volume will be the benchmark for all future discussions of the *Logical Syntax*.

Michael Beaney
September 2008

# Notes on Contributors

STEVE AWODEY is Professor in the Department of Philosophy, Carnegie Mellon University, USA. His research interests include category theory, logic, and the history of logic and analytic philosophy, and he recently co-edited (with E. Reck) *Frege's Lectures on Logic* and (with C. Klein) *Carnap Brought Home: The View from Jena*. He is among the editors of the *Collected Works of Rudolf Carnap*.

DENIS BONNAY is Associate Professor in the Department of Philosophy, University Paris X Nanterre, France. At Paris X, he works within the IREPh (Institut de recherches philosophiques). He is also a member of the IHPST (Institut d'histoire et de philosophie des sciences et des techniques) and of the Department of Cognitive Studies at the École normale supérieure in Paris. His main interests are in philosophical logic and in the interface between logic and linguistics.

JACQUES BOUVERESSE is Professor at the Collège de France and a member of the IHPST (Institut d'histoire et de philosophie des sciences et des techniques), France. His areas of research include Wittgenstein, philosophy of language, philosophy of logic, the Austrian philosophical tradition, and a general evaluation of contemporary philosophy.

A. W. CARUS is the author of numerous papers on Carnap, of a series of papers with Steve Awodey about Carnap and Gödel, as well as of the book *Carnap in Twentieth-Century Thought: Explication as Enlightenment*. He is among the editors of the *Collected Works of Rudolf Carnap*.

RICHARD CREATH is Professor in the Department of Philosophy and in the School of Life Science and Director Program in the History and Philosophy of Science, at Arizona State University, USA. He is the author of numerous articles on Carnap and Quine, the editor of *Dear Carnap, Dear Van*, the co-editor (with Michael Friedman) of *The Cambridge Companion to Carnap*, as well as the general editor of the *Collected Works of Rudolf Carnap*.

MICHAEL FRIEDMAN is Frederick P. Rehmus Family Professor of Humanities and Professor of Philosophy at Stanford University, USA. His books include volumes on space-time physics, Kant, logical positivism, and most recently *A Parting of the Ways* and *Dynamics of Reason*. He is co-editor of *Kant's Legacy in the Nineteenth Century*, of *The Cambridge Companion to Carnap*, and of the *Collected Works of Rudolf Carnap*.

WARREN GOLDFARB has been on the faculty of the Philosophy Department at Harvard University, USA, since 1975. He has published in mathematical logic on decision problems and proof theory, and on the development of analytic philosophy and the interface of logic and philosophy, particularly on Frege, Russell, Wittgenstein (early and late), Carnap, and Quine. He is among the editors of the *Collected Works of Rudolf Carnap*.

THOMAS RICKETTS is Professor of Philosophy at the University of Pittsburgh, USA. He is the author of numerous articles on the development of analytic philosophy, especially Frege, Russell, Wittgenstein, Carnap, and Quine. He is among the editors of the *Collected Works of Rudolf Carnap*.

PHILIPPE DE ROUILHAN is 'Directeur de Recherche' at the CNRS (Centre national de la recherche scientifique). He teaches at the Université Paris 1 Panthéon-Sorbonne, and is a member of the IHPST (Institut d'histoire et de philosophie des sciences et des techniques), France. His works belong to logic, formal ontology, philosophy of logic, philosophy of mathematics, and philosophy of language. He is currently working on hyperintensional logic and on truth and logical consequence.

THOMAS UEBEL is Professor of Philosophy, University of Manchester, UK. His research interests include general epistemology, philosophy of science, and the history of analytic philosophy. Besides many publications in these areas he is the author of *Empiricism at the Crossroads*.

PIERRE WAGNER has been on the faculty of the Philosophy Department, University Paris 1 Panthéon-Sorbonne since 1994. He is a member of the IHPST (Institut d'histoire et de philosophie des sciences et des techniques), France. His research interests include philosophy of logic, the history of analytic philosophy, and the history of the philosophy of science.

# Acknowledgements

This book originates in a seminar on Carnap's *Logical Syntax of Language* which took place from October 2003 to January 2005 in Paris, at the IHPST (Institut d'histoire et de philosophie des sciences et des techniques). This seminar was attended by Denis Bonnay, Jacques Bouveresse, Serge Bozon, Jean Mosconi, Fabrice Pataut, Philippe de Rouilhan, and me. When the seminar was over, I proposed to organize an international conference on *Logical Syntax* which would be the basis of a book on the same subject. The contributors to this volume gathered in Paris on 5–6 October 2005 for a meeting which took the form of a workshop in which each paper for this volume was discussed. The Paris conference in October 2005 was organized thanks to a grant from the French Ministry of Research (ACI 'TTT', grant 02 2 0552). The preparation of the manuscript has benefited from a grant from the ANR (National Research Agency, France, project 'Logiscience', grant ANR-07-BLAN-0010-01).

I am very grateful to the participants in the seminar on the *Logical Syntax* at the IHPST, to the participants in the Paris conference, and to two anonymous readers for Palgrave Macmillan who made valuable comments on the papers and on the project as a whole. For their remarks on a first draft of the introduction, I am grateful to Denis Bonnay, Delphine Chapuis-Schmitz, Thomas Uebel, and, above all, to André Carus who made many detailed comments on it and helped me to improve it in many ways. I also want to thank Michael Beaney for accepting this volume in his 'History of Analytic Philosophy' series, and the editor for philosophy publishing at Palgrave Macmillan who was very supportive in this enterprise.

Open Court Publishing Company generously gave permission to quote extensively from *The Logical Syntax of Language*. I am grateful to Lance Lugar and Brigitta Arden, from the Archives of Scientific Philosophy in Pittsburgh, and to Robert D. Montoya, from the Department of Special Collections, Charles E. Young Research, UCLA, for their assistance in obtaining permission to quote from unpublished material by Carnap.

I am grateful to Peggy Cardon, Jindřich Černý, Jacques Dubucs, Sandra Laugier, Paolo Mancosu, Elisabeth Nemeth, Jean-Jacques Rosat, Sandrine Souraya, Jana Šrubařová, Volker Turm-Nemeth, and Mark Van Atten, who have been helpful in different ways.

Pierre Wagner
Paris, October 2008

# Note on References

Citations are given by author, date, and page numbers, with the following exceptions. When the context allows it, the author's name is omitted. Wittgenstein's *Tractatus* is usually cited by aphorism numbers. Carnap's *Logical Syntax of Language* and a few other works are either cited by page numbers of by section numbers. In the latter case, the sign '§' is placed before the section number. In some cases, two dates are given, separated by slashes. The first date, sometimes followed by a letter, determines a unique entry in the combined bibliography. The second number either refers to a translation or to a later edition given within that same bibliography entry. When two dates are followed by only one page number (or by a set of page numbers), the page number (or the set of page numbers) refers to the translation or to the later edition mentioned in the bibliography entry determined by the first date. The abbreviation 'LSL' is often used in the text and in citations. It refers to the 1937 English translation of Carnap's *Logische Syntax der Sprache*. In citations, it is used as an abbreviation of 'Carnap 1934d/1937'. Unpublished sources from the Carnap collections of the Archives of Scientific Philosophy in Pittsburgh and from the Department of Special Collections, Charles E. Young Research Library, UCLA, are cited by document name. The document name is preceded by the abbreviation 'ASP' in the first case, and by the abbreviation 'UCLA' in the second case.

# Introduction

*Pierre Wagner*

## 1 *The Logical Syntax of Language* and contemporary philosophy

*The Logical Syntax of Language* (LSL) is indisputably one of the landmarks in the history of analytic philosophy. Indeed, this remarkable achievement by one of the most prominent members of the Vienna Circle contributed to the constitution of the analytic tradition, and may be considered as a paradigm of scientific philosophy. Epitomizing high standards of clarity and precision, it introduced a bold philosophical project grounded on the most recent results in logic, and it advanced a large part of the way toward its realization. This was not achieved through hasty generalizations over approximate understanding of a piece of scientific knowledge. Quite the contrary: Carnap not only had an astonishing scientific erudition, but also a deep understanding of modern logic and its implications, and the book was praised as a monument both in the history of philosophy, and in the history of logic. Karl Popper considered that 'if ever a history of the rational philosophy of the earlier half of this century should be written, this book ought to have a place in it second to none' (1963, p. 203). Alfred Tarski, to whom the modern concept of logical consequence is usually attributed, acknowledged that the first attempt to give a precise definition of this concept was due to the author of LSL (1936b, p. 5). Evert Beth, well-known among logicians for his definability theorem, made the following comment: 'I expect Carnap's *Logical Syntax* to remain one of the classics in logical literature and hence to be read and studied by many generations of future logicians' (1963, p. 482). Carnap himself presents his book as having a purpose in *logic* – an exposition of the method of logical syntax – but this method is then put at the heart of a project which clearly has a *philosophical* value. Carnap thus gives a concrete realization of Russell's idea that philosophical problems really are logical ones, an idea that LSL pushes to its limits. He also takes up the idea of philosophy as a collective task to be completed step by step through the work of many as opposed to the realization of a system expressing one author's worldview. Though LSL bears the inimitable hallmark of its author's genius, Carnap himself certainly tended to see it as one step taken in a collective endeavour.

As soon as it was published in 1934, and even in some cases before its publication, the book was read and commented on by some of the best logicians and philosophers of the twentieth century. Quine first met Carnap in 1933 when he was writing LSL, and he had the chance to read the book in its manuscript form. After his return to the United States, he gave three enthusiastic lectures on Carnap's philosophical conceptions at Harvard University in November 1934.[1] Carnap also discussed his project with Gödel and benefited from his comments before publication. Tarski's comments on the original German edition led to corrections (acknowledged as due to Tarski) in the English translation.

Though LSL is widely known as one of the building blocks of the analytic tradition in philosophy, we can easily find reasons why few have actually read it. The foreword to the book announces a replacement of philosophy by the logic of science, but we know that such a replacement did not take place. A few years later, Carnap himself gave up the strictures of the syntactical method which are painstakingly elaborated in the book, allowing room for semantic concepts he had explicitly excluded during the syntactical period. If the author changed his mind after the publication of LSL, shouldn't we leave it aside as well, and concentrate on Carnap's later works? It is easy for a present-day reader to wonder why she should bother going through all the technicalities which take up a large part of the book and go to the trouble of mastering the peculiarities of its vocabulary, notations, and conventions. We also know that Carnap's ideas have been thoroughly scrutinized and criticized from different angles by such well-known logicians and philosophers as Tarski, Beth, Kleene, Gödel, Schlick, and Quine, and that the demise of Carnap's project is frequently viewed as one of the commonplaces in the history of philosophy. It is thus no wonder that after being recognized as one of the masterpieces of logical empiricism in the thirties, LSL has been considered for several decades as more a monument of the past than a stimulating book for the present.

Though this opinion is still entrenched,[2] we now have solid grounds for challenging it. Such a jaundiced view of Carnap's work resulted for the most part from oversimplifications and misconceptions that recent commentaries have amended or refuted as inappropriate. For example, we now know that Carnap's so-called 'semantic turn' after 1935 should not be conceived as the giving up of the philosophical project he had set out in LSL but rather as an extension of his method so as to include both a syntactical approach and a semantic one (Creath 1991; Ricketts 1996; Carus 1999). Second, one of the most widely accepted interpretations of Carnap's philosophical programme as a continuation of traditional empiricism by means of modern logic – an interpretation going way back and given currency by Quine and Goodman on the basis of Carnap's *The Logical Structure of the World* – has been challenged by a great number of commentators who have traced back neo-Kantian origins in Carnap's philosophy (Richardson 1998; Friedman 1999b). Third, the longstanding opposition between Carnap and Quine on the issue of analyticity is now much better understood, especially since the publication of their correspondence, together with previously unpublished materials (Creath 1990). As a result, a much more balanced judgement about the nature of their dissent is now possible (Creath 2004; Ricketts, this volume; Friedman, this volume and 2008). Fourth, Carnap's philosophy of mathematics in LSL and its criticism by Gödel have been investigated and clarified in recent years and it has been argued that, had Gödel's critique been published in time (Gödel 1995), Carnap would not have been short of a convincing rejoinder (Goldfarb and Ricketts 1992; Awodey and Carus 2004). Far from having been turned into a petrified system by its critics, LSL raises issues which have begun to be mastered only recently, more than seventy years after its publication, and which do prove to be fruitful for today's philosophy.

Traditional views on LSL began to change in the eighties when some philosophers started reconsidering logical empiricism in general, and Carnap's works in particular. Several scholars then realized that many an argument that had been usually considered as a refutation of Carnap's philosophy were actually based on misunderstandings, and that much remained to be learned from a careful examination of LSL. Such a re-evaluation of this work had preconditions, and it took time to satisfy them. It presupposed, first, that the technical difficulties of the book – the logical results Carnap states, explains, and uses, as well as their consequences – were mastered and digested. Because some of Carnap's points are closely related to the state of logic in the early thirties, such mastery not only required knowledge in logic itself but also in its history. Secondly, it presupposed the examination of unpublished manuscripts and Carnap's correspondence. The excavating work scholars have done in the archives proved indispensable for a better understanding of Carnap's philosophy before 1931, of the reasons why his views changed radically at that time, and of his exact aims after he had passed through the stages that led him to the philosophical position we find in LSL. Much work has been done on Carnap's other publications, most notably on *The Logical Structure of the World* (often called the *Aufbau* after its German title), but also on his earlier papers, on his ideas on conventionalism, on his research in logic, on his reinterpretation of epistemology, on his philosophy of mathematics, and on his relations with other members of the Vienna Circle. The work thus accomplished by Carnap scholars helped to satisfy a third precondition for a general re-evaluation of LSL: getting rid of a prejudiced inclination against Carnap's thought.

After more than twenty years of Carnap studies, we now have a much better understanding of his philosophy in the twenties and the thirties, and although there remain points on which commentators do not agree, we are in a much better position to pave the way for an easier and more balanced reading of LSL. The purpose of this book is to review some of the most important issues in the interpretation of the LSL, on the basis of the most recent studies on Carnap's philosophy, many of which have actually been conducted by contributors to this volume. In this introduction, my attempt will be to remove the main obstacles that stand in the way of the non-specialist. I shall also provide a précis of the book, clarify distinctive features of Carnap's vocabulary and conventions, explain technical details which may cause trouble to the philosopher who is not a logician or to the logician who is not an historian of logic, provide indications about the situation of LSL in the evolution of the author's thought, point out the disputed questions among commentators, briefly touching on the history of its interpretation, and, most importantly, I shall highlight the main issues of the book.

## 2 Editions

Carnap's book was first published in 1934 in Vienna (by Julius Springer) under the title *Logische Syntax der Sprache* as the eighth volume of a series edited by Philipp Frank and Moritz Schlick.[3] The foreword is dated 'Prague, May 1934'. However, in the preface to the English edition, Carnap states that the manuscript of the German original was sent to the publisher in December 1933. A second unchanged edition of the German version appeared in 1968. The English translation, due to Amethe Smeaton, Countess von Zeppelin,[4] was published in 1937 in London (by Kegan Paul Trench, Trubner & Co) and in New York (by Harcourt, Brace & Co), and was printed again seven times between 1949 and 1971. After 1971, there has been no new printing for almost thirty years, and it is only since 2000 that the book is available again, which is telling of its destiny in the history of analytic philosophy. Today, it is published by Open Court and will appear, together with the German original, in the fifth volume of the *Collected Works of Rudolf Carnap*.

The German edition is divided into paragraphs numbered from 1 to 86. The English edition includes the translation of several sections which had to be omitted in the original manuscript because of lack of space. These additions have been made in such a way that the numbering of the paragraphs is consistent with the German edition: whereas §34 is replaced by §§34a–i, and §60 by §§60a–d, other paragraphs (16a, 38a–c, 71a–e) are inserted.[5] Other less important additions and corrections are indicated in the 'preface to the English edition'.

According to the foreword of LSL, Carnap's purpose in this book is to give a precise and systematic exposition of a new method for the syntactical study of languages. This method is first explained through its application to two specific languages named I and II, and then described in a general way so that it may be applied to any language. The upshot of this method is a new general programme for philosophy.

LSL is divided into five parts preceded by a foreword giving important indications on the general purpose of the book, and by an Introduction about the method of logical syntax. Part I consists in an exposition of Language I (the rules defining it are stated and explained) followed by remarks on the 'definite' form of language, of which Language I is an instance. Part II shows how the syntax of Language I may be formulated in Language I itself, using a technique due to Gödel. Part III gives an exposition of the 'indefinite' Language II followed by remarks on possible further developments of this language. Part IV is devoted to general syntax and constitutes the heart of the book: the preceding parts may be regarded as the preliminary exposition of the syntactical method in two specific cases before it is given in its full generality. In Part V, Carnap explains some consequences of the syntactical method: the replacement of philosophy by the logic of science, and, as a consequence, the reinterpretation of sentences of the 'material' mode of speech by sentences formulated in a 'formal' one.

In the mid-thirties, Carnap also published two simplified expositions of the method and of its impact on philosophy. The first one, that he himself calls a 'pamphlet' and a 'popular explanation of some ideas of the last chapter in [LSL]' in a letter to Quine (Creath 1990, p. 154), was published in 1934 as *Die Aufgabe der Wissenschaftslogik* (*The Task of the Logic of Science*) in the series 'Einheitswissenschaft' (unified science) edited by Otto Neurath. The second one is the written version of a series of three lectures that Carnap gave at the University of London in October 1934, and it was published in 1935 as *Philosophy and Logical Syntax*. These two booklets give a general idea of Carnap's philosophical views at that time but systematically avoid all technicalities. As a consequence, the reader cannot expect to get a deep understanding of the syntactical method in all its details from their sole reading. Neither of them explicitly states the principle of tolerance, a central idea of LSL. The same remarks hold for the paper 'On the Character of Philosophic Problems', another simplified exposition which appeared in 1934 in the first issue of the journal *Philosophy of Science*.[6]

Other papers belonging to Carnap's syntactical period give a complementary exposition of his philosophical views on some specific points. The following titles are probably the most important ones in this period:

- 'Formalwissenschaft und Realwissenschaft' ('Formal science and science of reality'), published in 1935, is the text of a lecture given in 1934 at the Preliminary Conference of the International Congresses for the Unity of Science in Prague.

- 'Von der Erkenntnistheorie zur Wissenschaftslogik' ('From the theory of knowledge to the logic of science'), published in 1936, is the written version of a lecture given in September 1935 at the First International Congress for the Unity of Science in Paris.[7]
- 'Testability and Meaning' is a most important book-length paper written in English and published in two parts (in 1936 and 1937) in *Philosophy of Science*. It deals with epistemological issues that are not discussed in LSL.

It is also worth mentioning 'Die physikalische Sprache als Universalsprache der Wissenschaft' ('The Physical Language as a Universal Language of Science'[8]), published in 1932, although this paper was written at a time when Carnap had not yet elaborated all the details of the method expounded in LSL and, most importantly, had not yet formulated the principle of tolerance.[9] Accordingly, it actually reflects an intermediary stage in the evolution of Carnap's thought: between the *Aufbau* and LSL. In this paper, and in another paper published earlier the same year, the project of LSL is mentioned under the name 'Metalogik' (1932a/1959, p. 78; 1932b, p. 435).[10]

## 3 Philosophy and the logic of science

In recent years, the extensive study of Carnap's notes, correspondence, and unpublished papers has enabled scholars to reconstruct the stages that led Carnap from his philosophical views in the late twenties to the ones that the reader finds in LSL. The story is told with great precision by Steve Awodey and André Carus,[11] who have managed to bring to light several layers in Carnap's philosophical views in LSL. For a long time, the exact connections between the *syntactical method* as a tool for philosophy on the one hand and the *principle of tolerance* on the other hand were not so clear. We now know that Carnap espoused these two ideas at different times and for different reasons. In LSL, although these elements of his philosophy are clearly distinguishable, Carnap neither confronts them with each other nor expounds them as separable. Only with hindsight do we know that soon after the publication of LSL, Carnap would relax the strictures of the syntactical method and adopt a broader one, whereas he would never give up the principle of tolerance in his later philosophy. More generally, commentators have made a great effort to articulate what exactly the new and lasting elements of Carnap's philosophy that surface in LSL are, trying to distinguish these elements from the ones which are specific of Carnap syntactical period and which he would not maintain in later works. From Carnap's point of view, philosophy was at a turning point, and this was a consequence of the revolution taking place in logic. For the new form that philosophy was to take, Carnap used the name: 'logic of science' (*Wissenschaftslogik*).[12] The latter happened to be temporarily linked with the syntactical method. In Carnap's later writings, the phrase 'logic of science' would generally be avoided, probably because it was too reminiscent of this very linkage.[13] But the main features of the idea would remain.

Here, I will first delineate some of the general characteristics of the 'logic of science' as they are depicted in LSL, and then highlight its relationships to the principle of tolerance. I shall thereafter focus more specifically on the method of logical syntax. To be sure, this is a retrospective distinction: in LSL, the logic of science is never disconnected from the syntactical method. Indeed, the former is *defined* as employing the latter.[14]

Following Russell and Wittgenstein, and like other members of the Vienna Circle,[15] Carnap maintained that logical analysis is an indispensable tool for the clarification of language. Its application to the sentences of traditional philosophy reveals some deeply entrenched illusions that deceive us, and which philosophers have often fallen prey to. While sharing the diagnosis, Carnap had his own ideas on the kind of cure that philosophy needs. Particularly damaging, according to him, is the confusion between *object-questions,* which pertain to some domain of objects, and *logical questions,* which are concerned with terms, sentences, theories, and other linguistic elements which refer to the objects in the domain under consideration (LSL, p. 277). Many problems of traditional philosophy which look like object-questions, Carnap maintained, are actually logical questions, and they should be treated as such. Philosophers are liable to such confusion, and they often entertain the illusion that they talk about things when logical analysis reveals that what they say concerns the form of language. This easily leads to pseudo-problems, talks at cross purpose, and endless disputes. One important aspect of Carnap's programme is to provide a cure for this kind of trouble. Thereby, all traditional philosophical problems are not systematically eliminated as such; but a more formal mode of speech is introduced, which prevents us from falling into some of the logical traps of word-languages. This mode of speech depends on a logical method which provides, essentially, a system of sharply defined concepts, to be used as tools for logical clarification (LSL, p. xiii). 'Analytic', 'synthetic', 'valid', 'contradictory', 'logical', 'consequence', 'derivable', 'equipollent', and 'synonymous' are typical examples of concepts belonging to this system.

At first sight, Carnap's philosophical programme in LSL sounds like a negation of philosophy: '*Philosophy is to be replaced by the logic of science* – that is to say, by the logical analysis of the concepts and sentences of the sciences' (LSL, p. xiii). Traditional philosophical problems are to be given up and replaced by a new agenda of questions, the purpose of which is to shed light on the logical relations among different parts of our discourse. Philosophy is not eliminated though.[16] First, because the logic of science may be considered with good reason as Carnap's specific way into the kind of scientific philosophy that was favoured in the Vienna Circle. Second, because Carnap, unlike Neurath, is not adamant on the rejection of the word 'philosophy' and he actually uses it.[17] Whether the term 'philosophy' should or should not be applied to the logic of science is for him 'a question of expedience' (LSL, p. 279). Third, because Carnap does mention or defend philosophical theses, in LSL and in other texts. The point is that philosophical theses have to be given a precise formulation which enables their integration into the logic of science. As a result, they lose the absolutist character they have in traditional philosophy and are relativized to some language (they may be relative either to some proposed language, or to a language actually in use, or to all languages...).[18] For example, in traditional philosophy, the thesis of the unity of science may be given a formulation like 'all the objects of science are of the same kind', which looks like an answer to an *object-question*, and which is assumed to be true or false independently of the language we intend to use. By contrast, a formulation like 'there is a language which enables the formulation of all our knowledge' includes a relativization to a language which makes it an answer to a *logical question*. Within the logic of science, a philosophical tenet like the thesis of the unity of science has the character of a linguistic thesis, not of an ontological one: '*The relativity of all philosophical theses in regard to language*, that is, the need of reference to one or several particular language-systems, is a very essential point to keep in mind' (Carnap 1935a, p. 78). In the logic of science as described in LSL, the philosophical theses are replaced by theses about the syntax of the language of science. After some remarks on 'physicalism' and 'unity of science', Carnap writes: 'It is easy to see that both are theses of the syntax of the language of science' (LSL, p. 320). The logic of science thus offers a clear and precise way of formulating philosophical theses through the exposition of the rules defining a language. In order to defend physicalism in the framework of the logic of science, a philosopher will not give ontological nor any other material-mode arguments; instead, he will provide the rules that define a physical language and he will show that 'every language of any sub-domain of science can be equipollently translated into the physical language' (LSL, p. 320).

Physicalism and the unity of science, which have been much discussed in the Vienna Circle, are neither presupposed nor directly argued for in LSL. But when Carnap chooses Language I and Language II as specific examples of languages, it is clear that the rules for these languages are carefully chosen so as to defend what may still be called 'philosophical theses' even though these theses lose the absolutist character they have in traditional philosophy. A precise statement of Carnap's motivation for choosing Language I and Language II will have to wait until more details have been given about these two languages and about the method of logical syntax, but we may already indicate one of the most central theses Language I and Language II are meant to support, a thesis about logic, mathematics, and analyticity.

It is frequently said that according to Carnap all true mathematical sentences are analytic. Yet, this is typically the kind of absolutist philosophical thesis Carnap criticizes and avoids by using the formal mode of speech. As a matter of fact, the reader will not find any such statement in LSL. 'Analytic' is a syntactical concept which has no absolute meaning: a sentence is said to be analytic only relatively to some language, either explicitly mentioned or left implicit when the context makes clear which language the sentence belongs to. Regarding the issue of analyticity, Carnap's specific choice of Language I and Language II is most significant. These languages are chosen among others because they satisfy properties expressed in a series of theorems, the most remarkable of which states that in both Language I and Language II, all the so-called 'logical sentences' are *determinate*, which means that they are either analytic or contradictory. This property is formulated in theorem 14.3 for Language I (LSL, p. 40) and in theorem 34e.11 for Language II (LSL, p. 116). The reason why these theorems are so remarkable is that on the basis of these languages, large parts of mathematics may be formalized as 'logical sentences' and derived from their defining rules. The fact that all 'logical sentences' are either analytic or contradictory is then a most unexpected completeness result in view of Gödel's first incompleteness theorem since Language I and Language II both include enough of arithmetic for Gödel's incompleteness theorems to apply.[19]

How is this completeness result philosophically relevant? To be sure, theorems 14.3 and 34e.11 do *not* prove that true mathematical sentences are analytic – Carnap would reject such an absolutist formulation anyway – and even less do they prove, of course, that true mathematical sentences are analytic in all languages – which is obviously false. What Carnap establishes with Language II – which encompasses a much larger part of mathematics than Language I – is rather that there exists a language which 'includes the whole of classical mathematics' (LSL, p. 83) and in which all true mathematical sentences may be formalized as *analytic sentences*. This is a typical example of the form a philosophical thesis may take in the framework of the logic of science. A philosophical absolutist thesis about mathematics ('mathematics is analytic') is replaced by a proved proposition about some particular language ('mathematics is analytic in Language II'). This is in complete agreement with the *principle of tolerance*, to which I now turn.

## 4 Logical pluralism and the principle of tolerance

The principle of tolerance is stated in several parts of LSL and finds its most explicit formulations in the foreword and in §17, each time in the context of remarks about the foundations of mathematics. In papers published a few years earlier – in 1930 and 1931 – Carnap was still defending a version of the logicist philosophy of mathematics according to which 'mathematics is a branch of logic'.[20] The position endorsed by Carnap in LSL does not only represent a move to a different thesis, it is rather a radical turn. In his book, he approaches the problem of the formalisation of science in a completely new way and this goes hand in hand with the idea of a replacement of philosophy by the logic of science.

Up to now, Carnap explains, researches in logic have been guided by the striving after 'correctness', and according to a widely held opinion, any language-form deviating from classical logic must be justified: 'the new language-form must be proved to be "correct" and to constitute a faithful rendering of "the true logic"' (LSL, p. xiv). Carnap's fundamental move in LSL is the rejection of this opinion:

> the view will be maintained that we have in every respect complete liberty with regard to the forms of language; that both the forms of construction for sentences and the rules of transformation [...] may be chosen quite arbitrarily. (LSL, p. xv)[21]

It is hard to exaggerate the significance of such a standpoint, which Carnap calls the 'Principle of Tolerance' and which finds its most often quoted formulation in §17: '*It is not our business to set up prohibitions, but to arrive at conventions*' (LSL, p. 51), a statement that Carnap clarifies a few paragraphs later in the following way:

> *In logic, there are no morals*. Everyone is at liberty to build up his own logic, i.e. his own form of language, as he wishes. All that is required of him is that, if he wishes to discuss it, he must state his methods clearly, and give syntactical rules instead of philosophical arguments. (LSL, p. 52)

Today, these passages of LSL are probably the most often quoted, commented, discussed, and criticized, no doubt because they introduce a radically new conception of logic and of its function within science and because they have far-reaching philosophical consequences. What we find here is probably the main key to understanding Carnap's project in LSL, and the large space devoted to its interpretation in the essays of this volume simply reflects the place it actually has in the literature about Carnap today.[22] Simple as it may seem, the principle of tolerance has a lot of implications that are not easy to grasp and it is no wonder that its interpretation has been so much disputed.

It is difficult to fully appreciate Carnap's motivation for this principle without placing it in the context of the contemporary debates on the foundations of mathematics, and without stating precisely how Carnap diverges on this issue from the very logicians who also had a strong positive influence on his thought: Frege, Russell, Hilbert, and Wittgenstein.[23] On the other hand, the significance of the principle of tolerance is not limited to the domain of logic and mathematics, and Carnap did not give it up after his syntactical period. The radical turn which has been taken with its adoption would never be repudiated in his later work:

> This neutral attitude toward the various philosophical forms of language, based on the principle that everyone is free to use the language most suited to his purpose, has remained the same throughout my life. It was formulated as 'principle of tolerance' in *Logical Syntax* and I still hold it today, e.g., with respect to the contemporary controversy about a nominalist or Platonic language. (1963a, p. 18)

For the time being, we will limit ourselves to a few general remarks on the meaning of this principle, especially in the philosophy of mathematics.

The first thing to note is that the principle of tolerance is not a thesis but an 'attitude' (LSL, p. 51 and p. 165) or a 'standpoint' which is 'suggested' (LSL, p. xv).[24] Carnap's proposal is the following: if someone wants to build her own logic, i.e. her own form of language, we shall not ask her to show that the revision she wants to introduce is justified or correct. We will ask for a clear statement of the rules which define her new form of language and then analyse and discuss the consequences which would result from its adoption. For this kind of analysis and discussion, we have powerful tools at our disposal: the syntactic concepts of the method of logical syntax, which is the special form that the logic of science takes in LSL. Whereas Carnap's stance in the debates about the foundation of mathematics was still some version of logicism in 1930 (although a quite different version from Frege's or Russell's), he adopts a completely different strategy in 1934: his research does not aim anymore at establishing a *foundation* of mathematics which could be termed 'true' or 'correct'. The point is that regarding logic or the framework of a language, there is nothing to be 'true' or 'correct' of, so that anyone is free to adopt one's own logic, one's own set of rules for the definition of a language. Carnap's proposal is to stop considering formalism, logicism, intuitionism... as philosophical theses in the traditional meaning of the term and to adopt *logical pluralism* so as to replace the endless debates between the proponents of these doctrines by the metalinguistic study (using the conceptual tools of the syntactical method) of the consequences that would result from the adoption of corresponding forms of language.

In the context of logical pluralism, what is Carnap's own proposal when it comes to the choice of a language for science? No doubt his preference is to adopt a language in which true mathematical sentences can be formalized as 'logical sentences' and such that these sentences can be proved to be analytic (this is, of course, only one of the properties that Carnap's preferred language will satisfy). Here again no argument of *correctness* is put forward in order to justify this choice. The point of the principle of tolerance is to consider only pragmatic arguments of *convenience*. So Carnap's reason for adopting such a language is not that mathematics as it stands *really is* analytic, but that the adoption of a language in which mathematical sentences are analytic will have a clarifying effect on the system of our knowledge. This is one of the main pragmatic grounds for the choice of Language II: Carnap proves that (in spite of Gödel's incompleteness theorem) it is perfectly possible to build a language which includes the whole of classical mathematics and enjoys some completeness property, with the consequence that true mathematical sentences may be formalized as analytic sentences of this language.[25]

## 5 Logical syntax

In LSL, the logic of science is implemented through the method of logical syntax:

> The book itself makes an attempt to provide, in the form of an exact syntactical method, the necessary tools for working out the problems of the logic of science. This is done in the first place by the formulation of the syntax of two particularly important types of language which we shall call, respectively, 'Language I' and 'Language II'. (LSL, pp. xiii–xiv)

Before expounding with more details the method of logical syntax, we need to clarify the two cardinal terms 'language' and 'syntax' since Language I and Language II are not languages in the ordinary sense, and in LSL, 'syntax' does not have its usual meaning either.

The distinctive features of Languages I and Language II reflect Carnap's concept of language in LSL. First, Carnap distinguishes *word-languages* (German, Esperanto...) and *symbolic languages* which use symbols (*Zeichen*) instead of words and enable exact formulations. Language I and Language II are of the latter kind. Second, Language I and Language II are regarded as *calculi* and their definition is purely formal. In the context of LSL, this means that nothing is assumed about the nature of the symbols of the alphabet and that the languages, as well as the categories of symbols and expressions, are defined by rules which make no reference to the meaning of these symbols and expressions. Third, Language I and Language II are not only defined by *rules of formation* telling us which sequences of symbols are to count as sentences but also by *rules of transformation* giving conditions for the deducibility of a sentence from other sentences. Accordingly, they would rather be called 'formal systems' nowadays. The term 'language-system' is actually used in some papers (e.g. Carnap 1935a, pp. 41ff.). Fourth, Carnap carefully distinguishes the logical and the descriptive symbols of a language. In the case of Language I and Language II, the set of logical symbols is defined explicitly, but the list of the descriptive symbols is left implicit. Those symbols are to be added according to the use we intend to make of the language considered. Suppose we want to express the temperature at different positions, we may add a descriptive symbol 'te' to be used as a function symbol, in such a way that the expression 'te(3) = 5' means 'the temperature at the position 3 is 5' (LSL, p. 14). Language I and Language II may vary according to the list of the descriptive symbols which are added. Therefore, they are properly speaking more like types or families of languages than like particular instances, though Carnap terms them 'languages'. A language may also have no descriptive symbol at all, in which case it is said to be *logical* (LSL, p. 178).

The quite specific form of the objects of investigation that Carnap calls 'languages' in LSL depends on the method that is to be applied to them in the book. In the fourth part, when Carnap is about to give an exposition of *general* syntax – the syntactical method as applicable to *any* language – he makes clear how he proposes to use the word 'language' in this context:

> By a language we mean here in general any sort of calculus, that is to say, a system of formation and transformation rules concerning what are called *expressions*, i.e. finite, ordered series of elements of any kind, namely, what are called *symbols*. [...] In what follows, we will deal only with languages which contain *no expressions dependent upon extra-linguistic factors*. [...] two sentences of the same wording will have the same character independently of where, when, or by whom they are spoken. (LSL, pp. 167–8)

This does not mean, however, that language reduces to a calculus, as §2 makes clear: as a 'historically given method of communication' (LSL, p. 5), language may be studied from various viewpoints (psychological, sociological, historical...) taking into consideration its relations to meaning, speakers, action, perception... but the syntactical method abstracts from all aspects of language that fall outside the definition of a calculus. It considers only the *formal* aspect and may be applied only if rules of formation and transformation can be provided. This is precisely what makes it *syntactical*. The adoption of such a specific viewpoint on language raises the important issues of the relationships between calculi and ordinary word-languages and, consequently, of the applicability of the syntactical method to the latter. The reasons Carnap gives for choosing languages like I and II rather than natural word-languages as examples for the application of the syntactical method are *practical* reasons:

> In consequence of the unsystematic and logically imperfect structure of the natural word-languages (such as German or Latin), the statement

> of their formal rules of formation and transformation would be so complicated that it would hardly be feasible in practice. (LSL, p. 2)

Here, Carnap seems to assume that any natural word-language may be regarded as a calculus.[26]

In LSL, the word 'syntax' also has a quite specific meaning. As Carnap points out in the introduction, *syntax* and *logic* are usually thought of as theories of different types. Whereas the syntax of a language studies the structure of this language independently of any consideration of meaning, truth, or deducibility, logic is supposed to formulate the rules according to which a conclusion may be inferred from premises. But Carnap's point is precisely that the usual methods of syntax may be extended and applied to logical questions as well. Indeed, Carnap wants to show that, unbelievable as it may seem, the means of syntax actually suffice to realize all that can be done in logic. In other words: he wants to show that logic itself is nothing but a kind of syntax of language. This naturally requires further clarification.

The first important point is that Carnap defines the logical syntax of a language as 'the formal theory of the linguistic forms of that language' (LSL, p. 1). Here, 'formal' has a precise if somewhat unusual meaning:

> A theory, a rule, a definition, or the like is to be called *formal* when no reference is made in it either to the meaning [*Bedeutung*] of the symbols (for example, the words) or to the sense [*Sinn*] of the expressions (e.g. the sentences), but simply and solely to the kinds and order of the symbols from which the expressions are constructed. (LSL, p. 1)

We may be tempted to see this definition as serving the purpose of excluding from logic all that has to do with semantics. However, the reader needs be cautious because on the one hand what we call 'semantics' today simply did not exist at the time Carnap was writing LSL and because on the other hand, as we will see, there are nevertheless good reasons to term certain uses of Carnap's formal syntactic method 'semantics in disguise'.

The second important point is that the method rests on a distinction between the language under investigation, or 'object-language,' and the language in which the investigation is conducted, or 'syntax-language' (which we would now call a metalanguage). Here, the term 'syntax-language' does *not* necessarily refer to a language that features only means of definition characteristic of the syntactic method – the syntax-languages Carnap uses, such as German or English, are not so restricted – but to the fact that the definitions of the concepts characteristic of Carnap's method use only a restricted (syntactic) part of this language. The nature of this restriction has been much debated in the literature and will require some discussion: it turns out to be of the utmost importance for the interpretation of LSL. Carnap usually takes as a syntax-language a natural word-language with additional specific symbols, but it would also be possible to use a symbolic language: 'The syntax language may be either a word-language or a symbol-language, or, again, a language composed of a mixture of words and symbols' (LSL, p. 154). This being admitted, the critical and disputed issue may then be formulated in the following way: what part of the expressive and demonstrative power of this language – especially when compared with the object-language – may be used if the strictures of the syntactical method are to be respected? What kind of restriction does the method of logical syntax impose on the definition of syntactical terms?

Clearly, the method requires that no class of expressions of the object-language be defined in the syntax-language on the basis of the *meaning* of symbols or expressions, or on the basis of the *truth values* of sentences of the object-language. This would indeed be in outright contradiction with the definition of 'formal' quoted above. This kind of restriction is reminiscent of the methods used in the Hilbert School whose favoured solution to the problem of the foundations of mathematics was often called 'formalist'. As a matter of fact, Hilbert's metamathematics is one of the major sources of inspiration for Carnap's idea of a metalinguistic analysis of language, a point that is underlined in the foreword (LSL, p. xvi).

Hilbert's method consisted in formalizing mathematical theories such as arithmetic or analysis, thus giving to mathematical proofs the form of sequences of sequences of signs (sequences of formulas), and in reasoning about the formal proofs thus obtained from a metamathematical point of view, without referring in any way to an *interpretation* of the symbols and formulas. Consequently, properties of the formal systems in which mathematical theories had been formalized were provable at the level of *meta*mathematics. Since the objects of a metamathematical reasoning had been reduced to elementary entities (finite sequences of finite sequences of signs, the meaning of which was disregarded), Hilbert thought that only elementary demonstrative means would be required, at the metamathematical level, to prove the desired properties (e.g. the consistency of the formal systems under consideration). The programme was of an epistemological nature: it aimed at proving properties about more and more powerful formalized mathematical theories (arithmetic, analysis, set theory...) using metamathematical means that were so elementary that they did not themselves require any further proof of non-contradiction. Because Hilbert's method consisted in metamathematical reasoning on *formal* expressions, it is sometimes called 'syntactical'. In this context, the word 'syntactical' is associated with the use of a weak metamathematical language. Now it is essential to resist such an association when 'syntactical' applies to Carnap's method in LSL. Carnap certainly takes up Hilbert's idea of reasoning about formal expressions with no reference to their meaning whatsoever. But he also departs from Hilbert's formalist methodology and extends it in two important ways. First, Carnap's syntactical method is much more than a meta*mathematics* since it encompasses languages for the formalization of science in general and not only formalized mathematical theories:

> Whereas Hilbert intended his metamathematics only for the special purpose of proving the consistency of a mathematical system formulated in the object-language, I aimed at the construction of a general theory of linguistic forms. (Carnap 1963a, p. 54)

Second, Carnap explains in §45 that the syntactical method is not necessarily restricted to 'definite' concepts, which are always decidable: it admits the use, in the syntax-language, of powerful means of demonstrations that were clearly excluded by Hilbert's metamathematics.[27]

The word 'syntax' in LSL should thus be understood both in comparison and in contradistinction to the same word as it is used in grammar, and as it is used when speaking about the Hilbert School. Understanding Carnap's unusual characterization of logic in terms of syntax of language requires not only that we conceive the rules of transformation as being part of the syntax of a language but also that we resist the temptation of identifying Carnap's syntactical method to what is often called 'syntax' in the history of logic when reference is made to Hilbert's programme.

Hilbert had good reasons for sticking to weak metamathematics in proofs of non-contradiction: weak methods of proof being considered safer than more powerful ones, it seems that there would have been no point in proving the consistency of a non-elementary mathematical theory T using in the metalanguage the very methods available in T or more powerful ones. However, Carnap *does* allow in the metalanguage non-elementary methods of proof that were excluded by Hilbert's metamathematics. This is exactly what happens in §34i, which contains a proof of non-contradiction for Language II. Here is Carnap's comment on the proof he has just given:

> Hilbert set himself the task of proving 'with finite means' *the non-contradictoriness of classical mathematics*. [...] The proof which we have just given of the non-contradictoriness of Language II, in which classical mathematics is included, by no means represents a solution of Hilbert's problem. Our proof is essentially dependent upon the use of such syntactical terms as 'analytic', which are indefinite to a high degree, and which, in addition, go beyond the resources at the disposal of Language II. [...] Since the proof is carried out in a syntax-language which has richer resources than Language II, we are in no wise guaranteed against the appearance of contradictions in the syntax-language, and thus in our proof. (LSL, p. 129)

If Carnap's purpose was to justify and to secure mathematical reasoning, his proof would be pointless since the resources of the syntax-language used in the proof are non-elementary and no less subject to possible contradictions than the resources of Language II itself. If this is so, what is the point of giving such a proof? What is its epistemological value? More generally, if Carnap does not aim at giving a solution to Hilbert's problem, what is the purpose of the method of logical syntax? Here, Carnap's distance from any kind of foundationalist epistemology becomes palpable. In the framework of the logic of science and in agreement with the principle of tolerance, Carnap states the defining rules of a proposed language – in this case, Language II – and uses concepts provided by the syntactical method in order to prove properties that result from the choice of these rules. The method of logical syntax provides tools for the logic of science in the context of the principle of tolerance: '[...] no question of justification arises at all, but only the question of the syntactical consequences to which one or other of the choices leads, including the question of non-contradiction' (LSL, p. xv).

At this point, a serious problem arises that was first put forward by Beth in 1937. Suppose a logician suggests using language $S_1$ whereas another logician suggests using language $S_2$.[28] If they both acknowledge the principle of tolerance, they will not try to prove that their own language is the correct one; they will rather examine the consequences of the defining rules of $S_1$ and $S_2$ and compare the respective properties of these object-languages. For this, they will use the conceptual tools provided by the syntactical method in the syntax-language. The following question then arises: which language should they use as a syntax-language? The crucial point is that the provable properties of $S_1$ and $S_2$ may depend on the choice of the syntax-language. For example, the possibility of a proof of non-contradiction for Language II depends on the resources available in the syntax-language, as Carnap remarks in §34i. Here, Beth perceives a gap in the syntactical method: 'the results of the syntactical analysis of [the object-language] O are relative to the syntax-language which is applied' (Beth 1937, p. 165). This cardinal issue has been much discussed in the recent literature. Michael Friedman once argued that Carnap was committed to a weak conception of logic in the metalanguage (Friedman 1988), but he later adopted a different view (Friedman 1999a, p. 230). In his contribution to this volume, he puts forward a new solution based on Carnap's answer to remarks made by Beth in the Schilpp volume (Beth 1963). In a paper which critically discusses Gödel's argument against Carnap,[29] Warren Goldfarb and Thomas Ricketts have argued that while Carnap is not committed to any weak or neutral metalanguage, the admission of a metalanguage stronger than the object-language nevertheless has consequences that actually conflict with his claim that mathematics is non-factual, for reasons different from those put forward by Gödel (Goldfarb and Ricketts 1992). This issue is also discussed in Ricketts (2007).

## 6 Logic

Any discussion of Carnap's syntactical method and its implications should take into consideration that LSL was written at a time when modern logic as it is usually understood and presented today was not yet settled in its main lines. Most contemporary logical textbooks convey a general view of logic which has become standard only in the second half of the twentieth century and the techniques and notations used by Carnap may sometimes seem unusual to the contemporary reader. First, some of the notations Carnap introduces happen not to be in use anymore. For example, in LSL, gothic letters are used in the metalanguage as names for categories of expressions of the object-language; e.g. '$\mathfrak{S}$' (the gothic letter for 'S') designates an arbitrary sentence (*Satz*) of the object-language; '$\mathfrak{zz}$' (where '$\mathfrak{z}$' is the gothic letter for 'z') designates any numeral (*Zahlzeichen*) of the object-language.[30] Second, some familiar features of today's logic depend on concepts and techniques that had not been discovered and were simply not available to Carnap at the time he was working on LSL. A striking example is the concept of an $\mathcal{L}$-structure (for a formal language $\mathcal{L}$) which is commonly used in today's semantics and that Carnap does not have in LSL. Third, some features of Carnap's logic which are uncommon today are essential to Carnap's philosophical point. The most obvious example is the concept 'analytic in language S'. Whereas according to Gödel's first incompleteness theorem any consistent system which contains some (quite elementary) part of arithmetic is (deductively) incomplete, Carnap wants all the so-called 'logical sentences' of Language I and Language II to be determinate (i.e. either analytic or contradictory). So the concept 'analytic' (in I or II) must be more encompassing than the concept 'provable' (in I or II). For this reason, these two languages are endowed with rules of transformation much stronger than the rules of inference one finds in today's standard systems of formal derivation. Of course, this way of circumventing Gödel's first incompleteness theorem comes at a price: whereas in both languages the set of provable formulas is recursively enumerable, the set of analytic sentences is not; a fact that Carnap expresses in his own way in terms of 'indefiniteness' (LSL, §34a).

### 6.1 Contemporary deductive logic

I shall first give a survey of the standard content of today's textbooks in logic in order to underline the main differences with LSL. Of course, there are many different ways of expounding today's basic deductive logic. I shall consider one standard way among others and skip details.[31]

A typical textbook starts with *propositional logic*, moves on to a more detailed study of *first-order logic*, and ends with remarks on *second-order logic*. In each case, the presentation of the syntactical characteristics of the language precedes the definition of *semantic* concepts.

The vocabulary of languages for *propositional logic* contains

i. parentheses and symbols for propositional connectives ($\neg$: negation, $\wedge$: conjunction, $\vee$: disjunction, $\rightarrow$: conditional...) as *logical symbols* and
ii. propositional letters ($p, q \ldots$) as *non-logical* symbols.

The vocabulary of languages for *first-order logic* contains

i. parentheses, comma, symbols for propositional connectives, individual variables ($x, y \ldots$) and quantifiers ($\forall$: for all, and $\exists$: there exists),
ii. individual constants ($a, b \ldots$) , first-order predicates ($M, N \ldots$) and first-order function symbols ($f, g \ldots$).

It usually also includes a symbol for equality.

A *formal language* $\mathcal{L}$ is a set of formulas defined by formation rules on the vocabulary. A *sentence* of $\mathcal{L}$ is defined as a closed formula, i.e. a formula in which any occurrence of an individual variable is in the range of a quantifier. One introduces some 'well-chosen'[32] method of formal derivation for first-order logic (typically: logical axioms and rules of inference) in order to define a notion of *formal derivability*: a formula $\phi$ is *formally derivable* from a set of formulas $\Gamma$ (notation: $\Gamma \vdash \phi$) if $\phi$ can be derived from formulas which are either logical axioms or formulas in $\Gamma$ using the rules of inference. A formula $\phi$ is a *formal theorem* (notation: $\vdash \phi$) if it can be derived from the logical axioms alone using the rules of inference. Whereas a *formal language* $\mathcal{L}$ is a set of formulas, a *formal system* is a formal language equipped with a method of formal derivation.

After these syntactical definitions, semantics begins when formal languages are interpreted, which is done in the general framework of set theory following techniques essentially due to Tarski. An *$\mathcal{L}$-structure* (or *$\mathcal{L}$-interpretation*) $\mathcal{I}$ is based on a non-empty set $\mathcal{D}$ – the *domain*, or *universe of discourse* – on which the non-logical symbols are interpreted by a function which associates:

i. one element of $\mathcal{D}$ to each individual constant;
ii. a subset of $\mathcal{D}^n$ (i.e. an $n$-ary relation) to each $n$-ary predicate;
iii. an $n$-ary function from $\mathcal{D}^n$ to $\mathcal{D}$ to each $n$-ary function symbol.

The interpretation of the logical symbols is standard and common to all the $\mathcal{L}$-structures.[33] An *assignment* $s$ is a function which associates an element of $\mathcal{D}$ to each individual variable. Concepts of Tarskian semantics are then defined:

i. *satisfaction* of a formula $\phi$ in a $\mathcal{L}$-structure $\mathcal{I}$ for an assignment $s$;
ii. *truth* of a sentence in a $\mathcal{L}$-structure;

iii. *universal validity* of a sentence: a sentence $\phi$ is *universally valid* (notation: $\models \phi$) if it is true in any $\mathcal{L}$-structure;
iv. *logical consequence*: a sentence $\phi$ is a *logical consequence* (or a *semantic consequence*) of a set of sentences $\Gamma$ (notation: $\Gamma \models \phi$) if $\phi$ is true in all the $\mathcal{L}$-structures in which all the sentences of $\Gamma$ are true.

The choice of the method of formal derivation (symbolized by the sign '$\vdash$') is crucial in the proof of the following central theorems which hold for any sentence *F* and any set of sentences *A*:

| | |
|---|---|
| (i) soundness of $\vdash$: | if $\Gamma \vdash \phi$ then $\Gamma \models \phi$ |
| (ii) Gödel's semantic completeness theorem: | if $\Gamma \models \phi$ then $\Gamma \vdash \phi$. |

All this gives a logical first-order framework in which a theory may now be formalized. A theory *T* couched in language $\mathcal{L}$ is a set of sentences of $\mathcal{L}$. Typically, one may want to formalize in $\mathcal{L}$ some version of arithmetic, of geometry, or of any mathematical or non-mathematical theory.

In standard textbooks, one usually does not find any extensive treatment of second-order logic. Any remarks which are made on this subject fall into two categories. First, second-order logic has a much greater expressive power than first-order logic. For example, equality becomes definable because it is now possible to quantify over properties: two individuals *x* and *y* are equal if, *for all properties P* of individuals, *x* has property *P* if and only if *y* has property *P*. Actually, in second-order logic, large parts of mathematics become provable on the basis of formulas which are 'purely logical', i.e. in which no non-logical symbols occur. And second, second-order logic does not enjoy semantic completeness: there exists no method of formal derivation $\vdash$ such that, for any formula *F* and any set of formulas $\Gamma$, if $\Gamma \models \phi$ then $\Gamma \vdash \phi$ (where the concept of universal validity symbolized by '$\models$' is supposed to have been defined appropriately for second-order languages). Some philosophers – most notably Quine – have used the conjunction of these two remarks to argue that second-order logic is no logic at all, but mathematics in the guise of logic. Such a viewpoint on the relationships between logic and mathematics, which is still controversial today, is to be contrasted with Carnap's, as will become clear shortly.

One also proves, or at least mentions, Gödel's first incompleteness theorem: if a consistent theory *T* couched in a language $\mathcal{L}$ (not necessarily first-order) contains some (quite elementary) part of arithmetic, then *T* is deductively incomplete, i.e. there exists a sentence *G* of $\mathcal{L}$ such that $T \nvdash G$ and $T \nvdash \neg G$. In other words, under the conditions just stated, there exists no method of formal derivation that defines a partition of the set of all the sentences of the language of theory *T* into sentences formally derivable from *T* on the one hand and sentences the negation of which is formally derivable from *T* on the other hand. One crucial point, here, with respect to LSL, is that the method of formal derivation symbolized by '⊢' is supposed to be 'effective' or, as Carnap would say, 'definite' (*definit*); in other words: there must exist a mechanical method which gives the answer 'yes' after a finite number of steps to the question 'is $\phi$ formally derivable from $T$?' whenever it is actually the case that $\phi$ is formally derivable from $T$. Carnap's concept of 'consequence' (*Folge*), although an essential tool of the 'syntactical' method, does not satisfy this property, and is therefore not termed a method of formal 'derivation' (*Ableitung*). It thus enables Carnap to obtain a partition of the 'logical sentences' of Language I and of Language II (which both contain enough of arithmetic for Gödel's incompleteness to apply) into analytic sentences on the one hand and contradictory sentences on the other hand.

Here, it begins to be palpable how remote Carnap's vocabulary is from today's standard terms in logic: 'language', 'formal', 'logical', 'consequence', 'syntax' and other cardinal terms just do not have the same meaning in our textbooks and in LSL.

## 6.2 Formal rules and material interpretation

In the definition of both Language I and Language II, Carnap considers in turn rules of formation (*Formbestimmungen*) and rules of transformation (*Umformungsbestimmungen*). The rules of formation are presented in a quite informal way: Carnap describes the main categories of symbols and expressions, gives examples, and underlines the intended meaning of symbols and the intended use of categories of expressions before providing definitions in a more formal manner. The informal comments by means of which Carnap makes his motivation explicit should not be confused with the formal application of the syntactical method.

While reading Carnap's clarifications about the meaning of symbols, we quickly encounter one major difference between his logic and ours. When a logician defines a formal language $\mathcal{L}$, he may justify his choice for non-logical symbols (individual constants, predicates...) by making explicit the intended $\mathcal{L}$-structure he has in mind. If he wants to formalize let us say arithmetic, he will probably equip language $\mathcal{L}$ with an individual constant to be interpreted by zero, a one-place function symbol to be interpreted by the successor function, a two-places function symbol for addition... and he will also make clear that the intended universe of discourse is the set of natural numbers. Yet, $\mathcal{L}$ is a non-interpreted language and there are an infinite number of $\mathcal{L}$-structures which, except in special cases, are not isomorphic to the intended interpretation. Such semantic considerations are to be contrasted with Carnap's syntactical method in which no concept of an $\mathcal{L}$-structure is defined. Though a syntactical definition must be 'formal' in so far as it refers only to 'the kinds and order of the symbols from which the expressions are constructed' (LSL, p. 1), this by no means implies that the symbols have no meaning and the expressions no sense. Language I and Language II are not 'formal' in *our* sense: they are *interpreted* languages. The point is that applying the syntactical method requires the interpretation to be disregarded, not the language to be a non-interpreted one. Even though Carnap considers the possibility of giving several interpretations to a calculus (LSL, p. 229) when be defines the concept 'interpretation of a language' in the context of general syntax (LSL, §62), he never resorts to any concept of an $\mathcal{L}$-structure. Throughout LSL, he mentions what he calls the 'material interpretation' (*inhaltliche Deutung*)[34] of the language he is considering, a clear indication that this language is interpreted. For example, after enumerating the primitive sentences of Language I, Carnap writes: 'we shall now see that all primitive sentences when materially interpreted [*bei inhaltlicher Deutung*] are true' (LSL, p. 30). The informal proof that follows, conducted in the metalanguage, is no application of the syntactical method. It is only an informal argument showing the usefulness of Language I as well as Carnap's motivation for choosing this language.

### 6.3 Logical symbols, logical rules, and mathematics

A fundamental distinction is made in LSL between *logical* symbols and non-logical or *descriptive* ones. We encounter here another difference between Carnap's logic and ours. First, in Language I and Language II, there is nothing like what we usually call 'individual constants', which serve the purpose of naming individuals in a specific universe of discourse. When materially interpreted, the expressions '0', '$0^{\mid}$', '$0^{\mid\mid}$', etc. are not names for objects but for positions: if '0' designates the first position, '$0^{\mid}$', '$0^{\mid\mid}$', etc. respectively designate the second position, the third position... (LSL, p. 13). Second, both '0' and '$^{\mid}$' are listed among the *logical* symbols and '$^{\mid}$' is not a function-symbol, although it has properties of a successor function in view of its material interpretation. These features of Language I and Language II are remarkable on several counts and must be contrasted both with today's usual presentations of logic and with Carnap's conception of logic in 1930, before his 'syntactical period'.

In today's standard textbooks, only *non-logical* symbols have zero and the successor function as their intended interpretation when a language is used to formalize some version of arithmetic. Though the issue of the bounds of logic is still much disputed, standard textbooks make a sharp distinction between a logical framework (a formal language $\mathcal{L}$ equipped with a method of formal derivation) and the formalization of a mathematical theory in this framework by non-logical axioms making essential use of non-logical symbols of $\mathcal{L}$. This is in sharp contrast with Language I and Language II where the list of the logical symbols is devised in such a way that mathematical sentences can be formalized using only logical symbols. What Carnap calls a 'logical sentence' ($\mathfrak{S}_{l}$ , with subscript 'l') is defined as a sentence in which only logical symbols occur (as opposed to a descriptive sentence $\mathfrak{S}_\mathfrak{d}$, with subscript 'd'). Moreover, axioms of arithmetic are included in the primitive sentences of both Language I (LSL, p. 30) and Language II (LSL, p. 92) and nevertheless considered as *logical* rules. Carnap's motivation is a conceptual clarification of the distinction between formal sciences (including logic and mathematics) and empirical sciences, and the syntactical definition of a 'logical sentence' is one step toward this clarification (to be followed by others in LSL). Languages I and II are devised in such a way that mathematics can be formalized using only logical symbols, and any 'logical sentence' is either analytic or contradictory. When Carnap remarks that Language II 'contains the whole of classical mathematics (functions with real and complex arguments; limiting values; the infinitesimal calculus; set theory)'[35] (LSL, p. 83), he does not mean that Language II constitutes a logical framework for the formulation of axiom systems for these theories.[36] Only *logical* symbols are required for the formulation of such axioms, and these axioms are formally derivable from the defining rules of Language II anyway. In §39, Carnap consider the special case of the real numbers:

> The usual axioms of the arithmetic of real numbers need not be set up here in the form of new primitive sentences. These axioms – and hence the theorems derivable from them – are *demonstrable* in Language II. (LSL, p. 148)

The inclusion of '0' and ''' among the logical symbols, and of axioms of arithmetic among the logical rules of Language I and Language II, should also be contrasted with Carnap's earlier view on logic and mathematics. It gives evidence of the evolution of his thought in the philosophy of mathematics. In several papers published in 1930 and in 1931,[37] Carnap defended the logicist idea according to which 'the concepts of mathematics can be derived from logical concepts through explicit definitions' and 'the theorems of mathematics can be derived from logical axioms through purely logical deduction' (1931a, p. 91). Clearly, when Carnap wrote this, he neither considered zero and the successor function as primitive logical concepts nor any sentence of arithmetic as a logical axiom. In 1930, he still followed Frege and Russell in taking the concept of logical symbol in a narrow sense and he took the trouble to explain his reasons for sticking to this philosophical position (Carnap 1930b). In 1934, when the principle of tolerance was applied, the decision to include mathematical symbols among the primitive symbols (as Hilbert had done) became a question of purely technical expedience. At that time, no philosophical argument could stop Carnap from counting '0' and ''' among the logical symbols any more. In so doing, he did not mean to abandon logicism but to reformulate it in a way that essentially eliminated the importance of any discussion of the distinction between logical symbols in the narrower sense and mathematical symbols.[38]

## 6.4 Logical consequence

If we compare Carnap's exposition of logic in his *Outline of Logic* (Carnap 1929) and in LSL, one of the most striking differences is the lack of any extensive treatment of the concept of logical consequence in the 1929 book and the centrality of this concept in LSL. In 1929, Carnap mentions two rules of inference (*Schlussregeln*) that he calls 'material principles' (*materiale Grundsätze*) and which, he explains,

> must be understood materially [*inhaltlich*] and cannot be expressed symbolically, because they give instructions on how to act, i.e. on how to assert new claims on the basis of claims already made. (Carnap 1929, p. 10)

This is reminiscent of Frege's universalist conception of logic, in which no concept of a metalanguage is ever considered: the inferences are carried out according to rules that cannot be formulated in the logical language itself.[39] After making the two rules explicit,[40] Carnap does not say anything more on this issue in his 1929 book. 'Derivation' and 'consequence' are not even mentioned in the index. This is in sharp contrast with LSL, where the definition of these concepts is one of the most central issues of the book. Here Carnap makes a most important distinction between derivation (*Ableitung*) and consequence (*Folge*). Again, a comparison with today's logic is revealing, though not easy, because our familiar distinction between formal derivation and logical (or semantic) consequence does not coincide with Carnap's, even if there are important relationships.[41]

Carnap's position regarding the two methods he calls 'methods of deduction' (*Deduktionsverfahren*) (LSL, p. 39 and p. 100) – namely a method of derivation (*Ableitung*), or d-method, and a method of consequence-series (*Folgereihe*), or c-method – must be understood in its historical context: as Carnap was working on LSL, he was well aware of Gödel's proof of the first incompleteness theorem (1931) but he did not know Tarski's work on the concept of truth.[42] Until 1931, logicians could still hope for a formal system with a consistent set of axioms out of which all the theorems of mathematics could be deduced through precisely stated rules of inference. Gödel's 1931 theorem of (deductive) incompleteness proves that these requirements cannot be satisfied by any extension of the system of Russell and Whitehead's *Principia Mathematica*. The possibility of equating 'mathematically true' with 'provable in (any extension of) *Principia Mathematica*' was thus refuted and many believed that, generally speaking, reconciling 'mathematically true' with 'provable in some formal system' was hopeless. Carnap himself notes that 'according to the more recent findings of Gödel, the search for a definite criterion of validity for the whole mathematical system seems to be a hopeless endeavour' (LSL, p. 99).

For Carnap, however, Gödel's theorem was, above all, a threat to the idea of a clear-cut distinction between the analytic sentences of formal science and the synthetic sentences of empirical science with no place left for any a priori synthetic sentences (an idea which was also shared by other members of the Vienna Circle). Consequently, Carnap responded to Gödel's incompleteness theorem with a new method of deduction – the method of *consequence* – that goes beyond the limits of any method of formal derivation. In the context of the definition of Language II, the difference is made clear in §34a: 'In the rules of inference, only a finite number of premises (usually only one or two) appear' (LSL, p. 99). By contrast, 'a method of deduction which depends upon indefinite individual steps, and in which the number of the premises need not be finite, we call a *method of consequence* or *c-method*' (LSL, p. 100). The introduction of a *c-method* enabled Carnap to vindicate completeness: 'In this way a *complete criterion of validity for mathematics* is obtained' (ibid.). But to achieve this result, Carnap had to give up the definiteness which characterizes the d-method: 'In order to attain completeness for our criterion we are thus forced to renounce definiteness, not only for the criterion itself but also for the individual steps of the deduction' (ibid.). Whereas what might be called Carnap's 'c-completeness' means that each logical sentence is either analytic or contradictory, it does not imply that each logical sentence is decidable (*entscheidbar*), i.e. either demonstrable or refutable. In §36, Carnap gives an important example of an undecidable logical sentence.[43]

Roughly speaking, Carnap's *method of derivation* is related to our method of formal derivation in so far as both methods are based on *finite* rules of deduction.[44] The main difference is that Carnap considers languages in which 'logical sentences' include mathematical sentences, whereas our contemporary methods of formal derivation are usually defined on the basis of logical signs in a narrower sense. In particular, for first-order languages, methods of formal derivation are chosen so as to enable a proof of the theorem of semantic completeness, a result which is foreign to Carnap's method of logical syntax.[45] As for the relationships between Carnap's concept of *consequence* and our concept of logical consequence, it might seem at first sight that there are not any: whereas Carnap's definition of consequence is given in the framework of a *syntactical* method, our concept of logical consequence is a *semantic* one, based on the concept of an $\mathcal{L}$-structure that Carnap does not have, and it is essentially due to a part of Tarski's work that Carnap was not aware of when he was working on LSL.[46] However, an examination of Carnap's way of elaborating his concept of consequence for Language II – based on a notion of 'evaluation' introduced in §34c – makes it clear that what he did there was, essentially, semantic in syntactical guise. The main difference between his concept of *evaluation* and Tarski's concept of *satisfaction* is that Carnap's evaluation consists in the replacement of (a possibly infinite number of) expressions by expressions whereas Tarski's relation of satisfaction holds between expressions and non-linguistic entities.[47] In his 1936 paper on the concept of logical consequence, Tarski acknowledged that Carnap was the first to attempt a precise definition of this concept and he consequently sketches a comparison of his own definition with the one Carnap gives in LSL (Tarski 1936b/1956, pp. 413–14 and pp. 416–18).

## 6.5 Truth

Before giving a syntactical definition of 'consequence in Language II' (LSL, §34f.), Carnap defines 'analytic in Language II' (LSL, §34d), a concept also based on the notion of evaluation introduced in §34c. At that point, he comes so close to a definition of truth that the reader may wonder why he did not actually give one. In the special case of a *logical* language (Language II with no descriptive symbols), Carnap's definition of 'analytic in II' *is* a definition of 'true in II'. But Language I and Language II are descriptive languages (LSL, p. 101 and pp. 181–2) with synthetic sentences. Carnap does not define any truth predicate in the context of LSL because a definition of '$\mathfrak{S}_1$ is true in S'[48] for a descriptive language S must also have recourse to the meaning of symbols that occur in $\mathfrak{S}_1$, or to a translation of $\mathfrak{S}_1$ in the metalanguage, and thus goes beyond the limits of the syntactical method.[49]

In the fourth part of LSL (on general syntax), when the concepts 'true' and 'false' are discussed in more detail, Carnap goes so far as to outline the procedure by which a predicate 'true in $S_1$' (for some object-language $S_1$) might be defined in a metalanguage $S_2$. He then adds the following remark: 'A theory of this kind formulated in the manner of a syntax would nevertheless not be a genuine syntax. *For truth and falsehood are not proper syntactical properties*' (LSL, p. 216). So far, everything is clear. But the next sentence – which seems to formulate Carnap's ground for excluding truth and falsehood from syntax – has struck many commentators as astonishing and misplaced: 'whether a sentence is true or false cannot generally be seen by its design, that is to say, by the kinds and serial order of its symbols'. The obvious objection is that whether a logical sentence is analytic or contradictory cannot generally be seen by its design either, although 'analytic in S' and 'contradictory in S' (for some language S) *are* syntactical properties. We may note that a similar objection cannot be raised against the formulation adopted by Carnap in *Philosophy and Logical Syntax* when the same issue is considered:

> We cannot define the terms 'true' and 'false' in syntax, because whether a given sentence is true or false will generally depend not only upon the syntactical form of the sentence, but also upon experience. (1935a, pp. 47–8)

So it might be conjectured that this is what Carnap actually had in mind in the former quotation. But commentators have usually tried to give a more literal interpretation of his surprising remark.

Some of them referred to it as they tried to understand what had prevented Carnap from taking the step that would have led him to a semantic definition of truth as early as 1934.[50] On this point, commentators disagree. According to Coffa, only a 'verificationist prejudice' prevented him from actually giving a syntactical definition of truth (Coffa 1987, pp. 567–8). Richard Creath wonders about Carnap's motivations for staying at all costs within the strictures of the syntactical method and he proposes the following interpretation: 'it is plausible to assume that Carnap was antecedently prejudiced against the concept of truth. [...] under the pernicious influence of Neurath, truth would have been called "metaphysical" and "absolutist" ' (Creath 1991, p. 411). Thomas Oberdan rejects both Coffa's and Creath's construals and argues that Carnap rejected the concept of truth because of his philosophy of language:

> what Carnap's argument purports to show is that the only consistent approach to the definition of truth entails a hierarchy of languages. And much of the attraction of the method of syntax [...] was that it provided the means for metalinguistic discourse without the hierarchy. (Oberdan 1992, p. 252)

To this interpretation, Thomas Ricketts objects that Carnap is completely aware of the fact that his conception of syntax commits him to a hierarchy of languages (Ricketts 1996, pp. 248–9). As a matter of fact, a consequence of Gödel's first incompleteness theorem for Carnap's systems (because they include large parts of arithmetic) is that '*everything mathematical can be formalized, but mathematics cannot be exhausted by one system*; it requires an infinite series of ever richer languages' (LSL, p. 222). Thomas Ricketts thus offers an alternative interpretation:

> Carnap rejects the notion of truth [in LSL] because, for good reasons, he believes the notion of truth to be both syntactically intractable and otiose in logic. Carnap's antipathy to truth is thus rooted more in technical than in philosophical considerations. (Ricketts 1996, p. 233)

Any attempt to interpret the rejection of the concept of truth in LSL should take into account the fact that Carnap did admit it as a central concept of logic both before and after the syntactical period, in each case on different grounds. In the pre-syntactical period, ' "true" and "false" are undefinable basic concepts' (1929, p. 3).[51] Here, Carnap follows Frege's idea of the undefinability of the concepts of truth and falsehood without in any way excluding them from logic. In the mid-thirties, when Tarski explained to him his technique for a definition of truth, Carnap immediately adopted it and this was the beginning of his semantic turn. Here is how Carnap relates this event in his 'Intellectual Autobiography':

> When Tarski told me for the first time that he had constructed a definition of truth, I assumed that he had in mind a syntactical definition of logical proof or provability. I was surprised when he said that he meant truth in the customary sense, including contingent factual truth. Since I was thinking only in terms of a syntactical metalanguage, I wondered how it was possible to state the truth-condition for a simple sentence like 'this table is black'. Tarski replied: 'This is simple; the sentence "this table is black" is true if and only if this table is black.' [...] When I met Tarski again in Vienna in the spring of 1935, I urged him to deliver a paper on semantics and on his definition of truth at the International Congress for Scientific Philosophy to be held in Paris in September. (1963a, pp. 60–1)

Again, it should be stressed that Carnap's so-called 'semantic turn' was not as sharp as one might think and should rather be viewed as an extension of the methods of LSL. In his *Foundations of Logic and Mathematics* (1939) and in his *Introduction to Semantics* (1942), Carnap does not replace the syntactical approach with a semantic one: he complements the syntactical calculi with semantic systems. Though views to the contrary have been expressed in the past, it is now 'generally agreed among commentators', André Carus writes, that 'the acceptance of semantics was not a fundamental discontinuity in Carnap's development' (1999, p. 20).

## 7 Language I

In the section 'The Foundations of Mathematics' of his 'Intellectual Autobiography', Carnap notes that 'the constructivist and finitist tendencies of Brouwer's thinking appealed to [the members of the Vienna Circle] greatly' and that he 'had a strong inclination toward a constructivist conception' (1963a, p. 49). This was the main motivation for defining Language I, which 'fulfilled the essential requirements of constructivism' (ibid.).[52] In the early thirties, the terms 'finitism', 'constructivism', and 'intuitionism' were not always clearly distinguishable.[53] So Language I can only pretend to 'realize' them 'in a certain sense' (LSL, p. 46) and may be seen as Carnap's own proposal for giving a precise meaning to these terms. In the framework of the logic of science, this is exactly the kind of task the syntactical method aims at: to replace a vaguely defined idea by a set of formal (formation and transformation) rules in order to achieve exactness in the discussion. The fact that the intuitionistic concept of the continuum cannot be formalized in Language I suggests that Brouwer himself would not have found Language I very appealing. More generally, his philosophy of mathematics was clearly incompatible with Carnap's idea of the logic of science, so that he would not have accepted this idea in the first place. Brouwer insisted on the open character of mathematics as a mental activity and unlike Carnap, he never held that 'the problems dealt with by Intuitionism can be exactly formulated only by the means of the construction of a calculus' (LSL, p. 46).

The finitist character of Language I is reflected in the fact that all the arithmetical predicates which can be defined in it are decidable and all the arithmetical functions which can be defined in it are computable. This is one of the main properties of what Carnap calls a 'definite' language (LSL, §15). In such a language, the definability of predicates (i.e. relation symbols) and 'functors' (i.e. function symbols) is restricted by the fact that only bounded operators are used in their definition, or in the definition of the expressions used in their definition. So Language I, in which all variables are *numerical* ($\mathfrak{z}$, *Zahlvariablen*) and all individual constants are *numerals* ('0', '$0^{|}$', '$0^{||}$'... abbreviated as '0', '1', '2'...), has only *bounded* universal and existential quantifiers, and the only way to express unlimited universality is the use of free variables (LSL, p. 21). A 'least number' operator (K-operator) is introduced but also bounded: (K $\mathfrak{z}_1$)$\mathfrak{Z}_1$($\mathfrak{S}_1$) means: the smallest number $\mathfrak{z}_1$ *up to number* $\mathfrak{Z}_1$ such that sentence $\mathfrak{S}_1$ is true, and if no such number exists, 0 (LSL, pp. 22–3).[54] Basically, the logico-mathematical part of Language I is limited to what we would call 'primitive recursive arithmetic'.

Language I is extensional. Although the propositional connectives are syntactically defined by formal rules, they have a 'material interpretation' which is given by truth tables as in classical logic. The notations used are '$\sim$' for negation, '$\vee$' for disjunction, '$\bullet$' for conjunction, '$\supset$' for implication, and '=' for equivalence. The equality sign is used both between numerical expressions (the material interpretation of $\mathfrak{Z}_1 = \mathfrak{Z}_2$ is that $\mathfrak{Z}_1$ and $\mathfrak{Z}_2$ designate the same number) and between sentences, but according to a notational convention which aims to make the reading easier, the symbol '=' is replaced by '$\equiv$' when it occurs between sentences. The primitive *logical* symbols of Language I are '(', ')', ',', '|', '$\sim$', '$\vee$', '$\bullet$', '$\supset$', '=', '$\exists$', 'K', '0', and the numerical variables ($\mathfrak{z}$). Primitive *descriptive* symbols are added, which are either predicates ($\mathfrak{pr}_b$) or functors ($\mathfrak{fu}_b$) of any finite number of places. Other numerals ($\mathfrak{zz}$), predicates ($\mathfrak{pr}$), or functors ($\mathfrak{fu}$) can be introduced through definitions. Whereas numerals and predicates may be defined only by explicit definitions, functors may be defined either by explicit or by recursive definition (LSL, §8).[55]

'Derivation [*Ableitung*] of a sentence $\mathfrak{S}_n$ with premisses $\mathfrak{S}_1 \ldots \mathfrak{S}_m$ in Language I' is defined on the basis of primitive sentences (*Grundsätze*) and of rules of inference (*Schlußregeln*) (LSL, p. 29). Primitive sentences are not given directly but through schemata, because no sentential variables are available in Language I. Schemata for the sentential calculus, the bounded quantifiers, identity, arithmetic, and the K-operator are laid down in §11 (LSL, p. 30) in such a way that mathematical primitive sentences are not distinguished from the ones *we* would call 'logical'. Among the four rules of inference listed in §12 is a version of the principle of complete induction (LSL, p. 32), so here again, mathematics is not distinguished from what *we* would call 'logic'. 'A derivation without premises is called a *proof* [*Beweis*]', and 'the final sentence of a proof is called a *demonstrable* sentence [*beweisbarer Satz*]' (LSL, p. 29). A sentence

> $\mathfrak{S}_1$ is called refutable [*widerlegbar*] when at least one sentence $\sim\mathfrak{S}_2$ is demonstrable, $\mathfrak{S}_2$ being obtained from $\mathfrak{S}_1$ by the substitution of any accented expression [i.e. 0, or $0^{\mathrm{I}}$, or $0^{\mathrm{II}}$, or $0^{\mathrm{III}}$...] for all the [numerical variables] $\mathfrak{z}$ which occur as free variables. (LSL, p. 28)

The key step in the definition of Language I is taken in §14 when Carnap introduces the concept of consequence (*Folge*) by adding a rule of transformation with an infinite number of premises (LSL, p. 38). This concept is the basis on which essential tools of the method of logical syntax are defined: not only such concepts as analytic (*analytisch*), contradictory (*kontradiktorisch*), and synthetic (*synthetisch*) (in Language I) but also such syntactical concepts as the logical content (*Gehalt*) of a sentence, equipollent (*gehaltgleich*), or synonymous (*synonym*) (in Language I).

> A sentence $\mathfrak{S}_1$ is called *analytic* (in I) when it is the consequence of the null class of sentences [...]; it is called *contradictory* when every sentence is the consequence of $\mathfrak{S}_1$; [...] it is called *synthetic* when it is neither analytic nor contradictory. (LSL, pp. 39–40)

Carnap states (without giving a proof) that some analytic sentences are not provable (theorem 14.2) and he then proves (theorem 14.3) that every logical sentence ($\mathfrak{S}_{\mathrm{I}}$) is either analytic or contradictory. Whereas the first result states that some true mathematical sentences which can be formalized in Language I are nevertheless not provable, the second one aims at giving a formal counterpart to the distinction between truth and falsity for those mathematical sentences which can be formalized in Language I.

What we have here is an illustration of Carnap's own way into scientific philosophy and into the use of logical analysis: his aim is 'to provide a system of concepts' (syntactical concepts such as analytic, synthetic, logical content, derivable, consequence, etc.) in order to arrive at an exact formulation of 'the results of logical analysis' (LSL, p. xiii). Here, clarification of concepts is not obtained through a reform of our ordinary word-language but rather through the construction of a formally defined language satisfying certain properties *and* laid down as a proposal for the language of science (but certainly not as *the correct* language). 'By means of the concept "analytic", an exact understanding of what is usually designated as "logically valid" or "true on logical grounds" is achieved' (LSL, p. 41). The definitions of 'analytic in Language I' and 'analytic in Language II' are two different proposals for capturing in a formal and exact way the set of those sentences the truth of which does not depend on empirical facts:

> In material interpretation, an analytic sentence is absolutely true whatever the empirical facts may be. Hence, it does not state anything about facts. On the other hand, a contradictory sentence states too much to be capable of being true; for from a contradictory sentence each fact as well as its opposite can be deduced. A synthetic sentence is sometimes true – namely, when certain facts exist – and sometimes false; hence it says something as to what facts exists. *Synthetic sentences* [*synthetischen Sätze*] are the *genuine statements about reality* [*Wirklichkeitsaussagen*]. (LSL, p. 41)[56]

The most important point in this quotation is Carnap's warning: 'in material interpretation'. The first sentence is *not* a definition of 'analytic sentence' and the following sentences should not be taken literally either. The purpose of Language I, Language II, and the syntactical method is to provide tools for transforming such misleading and inexact formulations of the material mode of speech (*inhaltliche Redeweise*) into exact ones in the formal mode of speech.[57]

## 8 Language II

Language II is an extension of Language I such that all sentences of the latter are also sentences of the former. In addition to the numerical variables (𝔷), Language II has predicate-variables (𝔭, to be read 'p'), functor-variables (𝔣, to be read 'f') as well as sentential variables (𝔰, to be read 's'). All these variables may be quantified, universally or existentially, and both quantifiers and the K-operator may be used with no bound. In addition to the predicates (𝔓𝔯, to be read 'pr') and the functors (𝔣𝔲, to be read 'fu'), there are predicate-expressions (𝔓𝔯, to be read 'Pr') and functor-expressions (𝔉𝔲, to be read 'Fu'). The symbol of identity is used between 𝔓𝔯 and 𝔉𝔲 as well as between ℨ (numerical expressions) and 𝔖 (sentences). Sentential symbols (𝔰𝔞, 'to be read 'sa') may be either variables or constants. A system of types is defined for 𝔓𝔯, 𝔉𝔲, ℨ, and 𝔄𝔯𝔤 (argument-expressions, to be read 'Arg'). Symbols, types, and formation rules for numerical expressions, sentences, and definitions in Language II are explained in §§26–29 (LSL, pp. 81–90). The result is a language with a much greater expressive power than Language I.

The primitive sentences of Language II (LSL, §30) include some version of the principle of complete induction, of the axiom of choice as well as axioms of extensionality. Two rules of inference (modus ponens and universal generalization) and definitions for d-concepts (derivation, proof, demonstrable, refutation...) are given in §31.

As in the case of Language I, the key step for Language II is the definition of the c-concepts (consequence, analytic, contradictory, synthetic...). But in the case of Language II defining 'consequence' is a quite complicated matter. The definition of 'consequence' for Language I is based on the so-called $\omega$-rule, or 'rule of infinite induction' (LSL, p. 173), according to which a formula $\mathfrak{S}_1$ with one free numerical variable $\mathfrak{z}_1$ may be inferred from the infinite class of formulas obtained by substituting 0 for $\mathfrak{z}_1$ in $\mathfrak{S}_1$, $0^{|}$ for $\mathfrak{z}_1$ in $\mathfrak{S}_1$, $0^{||}$ for $\mathfrak{z}_1$ in $\mathfrak{S}_1$... In material interpretation, the rule means that if the property expressed by $\mathfrak{S}_1$ is true of each natural number, then it is true of all. In Language II, the free variable in $\mathfrak{S}_1$ is not necessarily a numerical variable; it may be, for instance, a predicate-variable. Now, if the syntax-language is strong enough, there are numerical properties definable in the syntax-language which are not definable in Language II. So, in material interpretation, the property expressed by $\mathfrak{S}_1$ may be true of each property definable in Language II and not true of all the properties definable in the syntax-language. For this reason, a straightforward generalization of the $\omega$-rule would not have the desired results and the definition of 'analytic in Language II' proceeds in a completely different way.

In the original German version of LSL, Carnap gives only a summary of the whole issue in a single paragraph (§34), leaving the details for another publication which was to appear the following year as (1935d). In the English translation, §§34a–i replace §34. In this introduction, I can hardly do more than enumerate the main issues as stake in this extremely rich part of the book which has been much discussed in the literature.[58]

In §34a, Carnap discusses the two methods of deduction (derivation and consequence), and clarifies the requirements to be met in order to fulfil the task of giving a *complete* criterion of validity for mathematics, in spite of Gödel's incompleteness theorem. For completing this task, §§34b–c introduce the technique of *evaluation* of a sentence, which is the syntactical forerunner of Tarski's semantic method. This technique is the basis of the quite involved definition of 'analytic in Language II' given in §34d. Here, an interesting philosophical question arises because we need to quantify over valuations ('every valuation of $\mathfrak{v}_1$', p. 111) in the definition of analyticity, which means – in material formulation – that we have to consider *all syntactical properties* (for a given type) for the evaluation of some sentences. At this point, Carnap foresees possible objections: are we committed to

> the conception that the totality of all properties, which is non-denumerable and therefore can never be exhausted by definitions, is something which subsists in itself, independent of all construction and definition? (LSL, p. 114)

Carnap states the question, dismisses it immediately (on grounds of its metaphysical character and because it is typical of the material mode of speech), and replaces it by a syntactical one: 'can the phrase "for all properties..." [...] be formulated in the symbolic syntax-language?' The answer is affirmative. But the concept 'analytic in Language II' thus obtained is then relative to the syntax-language in which it is formulated and interpreted.[59]

The 'c-completeness' theorem is proved in §34e, and other syntactical terms (consequence, compatible, independent, content...) are defined in §§34f–g where a series of theorems related to these terms are also stated. Finally Carnap proves three striking results: both the principle of induction and the axiom of choice are analytic, and Language II is non-contradictory (§§34h–i). The (quite limited) epistemological significance of these results is discussed in connection with the principle of tolerance (LSL, p. 121, pp. 123–4, pp. 128–9). In order to prove that the sentence which formulates the principle of induction in Language II is analytic in Language II, Carnap uses the principle of induction at the level of the *syntax-language*. Carnap then argues that the proof is neither circular (because the syntax-language is not Language II) nor absolute (because it depends on the richness of the syntax-language). The same holds for the axiom of choice.

In §35, Carnap states the famous self-referential lemma in its general form: 'For every syntactical property, it is possible so to construct a sentence that it attributes to itself – whether rightly or wrongly – just this property' (LSL, p. 129) and he uses it to construct an undecidable sentence which is the analogue in Language II to the sentence constructed by Gödel in the proof of his first incompleteness theorem. In a footnote added to the 1965 edition of his Princeton Lectures, Gödel credits the general self-reference lemma to Carnap (Gödel 1934/1965, p. 63).

The final paragraphs of Part III (§§37–40) give an outline of some further possible developments of Language II. They suggest ways of treating classes, cardinal numbers, real numbers, and physical concepts in the framework of this language. This clearly aims at showing some of the advantages of using a language such as Language II as a framework for science. However, the purpose of LSL is not to decide such a general issue, so Carnap uses more cautious formulations and he does not go so far as to present Language II as his own proposal for the form of a language for science:

> It follows from all these suggestions that *all the sentences of physics can be formulated in a language of the form of II*. [...] According to the thesis of *Physicalism*, which will be stated later (p. 320) but which will not be established [*begründed*] in this book, all terms of science, including those of psychology and the social science, can be reduced to terms of the physical sciences. [...] For anyone who takes the point of view of Physicalism, it follows that our Language II forms a complete syntactical framework for science. (LSL, pp. 150–1)

## 9 General syntax

In Part IV, Carnap undertakes the construction of a *general* syntax, one which may be applied to any language. Constructing the logical syntax of a language S, in Carnap's vocabulary, consists in giving, in a syntax-language $S_1$, an exact definition of a series of terms related to S (consequence, content, derivable, analytic...), using only restricted means of $S_1$, the ones which are characteristic of the *syntactical* method. In the case of specific languages such as I or II, one starts with a description of some specific features of the language: a list of symbols, a list of formation and transformation rules, an enumeration of the symbols which are to count as logical, an explicit distinction between sorts of variables (numerical, sentential, predicate-variables...), the designation of a symbol, if any, which will be materially interpreted by zero, etc. In the case of general syntax, we do not know anything about the syntactical categories of the language under consideration, indeed general syntax must apply to any language whatsoever, including languages with no variables, no symbol for zero, no symbol for negation... This means that all syntactical categories (variables, connectives, logical symbols, quantifiers, natural numbers...) have to be *defined* according to the strictures of the syntactical method. The only basis for these definitions and the only presupposition is a

> definition of 'direct consequence' to be stated in the following form: "$\mathfrak{A}_1$[60] is called a direct consequence of $\mathfrak{K}_1$[61] in S if: (1) $\mathfrak{A}_1$ and every expression of $\mathfrak{K}_1$ has one of the following forms: ... ; and (2) $\mathfrak{A}_1$ and $\mathfrak{K}_1$ fulfil one of the following conditions: ..." The definition thus contains under (1) the formation rules and under (2) the transformation rules of S. (LSL, p. 169)

So the extremely challenging task of Part IV may be characterized as follows: 'show how *the most important syntactical concepts can be defined by means of the term "direct consequence"* ' (LSL, p. 168).

The first step consists in giving the distinctive feature of d-rules as opposed to c-rules and this is done in §47 (LSL, p. 171). What follows is a series of definitions for d-terms (derivable, proof, demonstrable, refutable, decidable...) in §47 and for c-terms (consequence, valid, contravalid, determinate, incompatible, dependent, content, equipollent, synonymous...) in §§48–9. For example, valid sentences are defined as consequences of the null class.[62]

A crucial step is taken in §50 where Carnap attempts nothing less than a general definition of 'logical'. We now realize that this is an enormous task on which logicians have worked hard to this day and which is still much discussed in the philosophy of logic. Carnap's motivation here is to achieve an exact understanding of the informal distinction between logico-mathematical truths conceived as depending only on the rules of language and the empirical truths depending on extra-linguistic factors. The syntactical definition of 'logical expression' tries to capture the idea that the distinctive feature of logical symbols and expressions is that 'each sentence constructed solely from them is determinate' (LSL, p. 177). In other words: the truth-value of each sentence constructed solely from them is determined by the rules of the language. It is then no wonder that the central theorem of c-completeness stated in the same paragraph ('every logical sentence is determinate') follows directly from the proposed definition of a 'logical' expression. This is in sharp contrast with the theorem of c-completeness for Language II which had to be proved on the basis of quite involved definitions and a series of theorems (LSL, p. 116). In the following years, Carnap renounced the definition given in §50 because he thought 'logical' would rather be defined on another basis. In his *Introduction to Semantics*, he discusses the modifications that the views explained in LSL have to undergo and writes:

> The most important change concerns the distinction between logical and descriptive signs, and the related distinction between *logical and factual truth*. It seems to me at present that these distinctions have to be made primarily in semantics, not in syntax. (1942, p. 247)[63]

Another most important distinction is made in the next paragraph (§51), between *logical rules* (L-rules) and *physical rules* (P-rules). Both Language I and Language II are based only on primitive sentences and transformation rules which, in material interpretation, have a logico-mathematical meaning. Now Carnap also considers languages defined by rules which, in material interpretation, have an extra-logical meaning. In agreement with the principle of tolerance, the decision to include P-rules is declared to be 'a matter of convention and hence, at most, a question of expedience' (LSL, p. 180).[64]

Many commentators deemed that Carnap was thus giving too much extension to the principle of tolerance and strongly objected to P-rules. One of the first to criticize Carnap on this point was Schlick, who argued that the laws of nature should not be confused with conventions.[65] The main issue, here, is to know whether the introduction of P-rules is blurring the distinction between empirical and non-empirical truths, a distinction which is of fundamental significance for Carnap and for other members of the Vienna Circle.[66]

Formally, the concept of *P-rules* results from the syntactical definitions of 'L-consequence' and 'P-consequence' given in §51 (LSL, p. 181), which are the basis for further syntactical definitions (L-language, P-language, L-content, P-content...), the most important of which is probably the definition of 'analytic'. An analytic (or L-valid) sentence is defined as an L-consequence of the null class (whereas valid sentences are consequences of the null class, and P-valid sentences are valid sentences which are not L-valid). Intuitively, the difference between P-valid and L-valid sentences is that an L-valid sentence remains valid under all possible substitutions of descriptive expressions. It is hard to overestimate the importance of these paragraphs (§§50–2) which are not only the fulcrum of the life-long debate between Carnap and Quine on the issue of analyticity,[67] but also of crucial importance for the interpretation of Carnap's philosophical programme in LSL.[68]

One of the most challenging tasks of the book (from a technical point of view) is undertaken in the following paragraphs (§§53–8) where Carnap offers syntactical definitions of terms such as 'predicate', 'functor', 'variable', 'sentential function', 'universal operator', 'existential operator', 'connective', 'negation', 'conjunction', 'numerical expression', 'numeral', 'arithmetic', and 'real number'. This is done in the framework of general syntax, i.e. using only syntactical means and on the sole basis of the concept 'direct consequence'. Carnap's ability to manipulate syntactical techniques becomes fascinating when it is applied to the definition of concepts which, to our eyes, are obviously of a semantic nature. This is the case for the concept of range (*Spielraum*) defined in §56. In material interpretation, the *range* of a sentence $\mathfrak{S}_1$ in language S is 'the class of all the possible cases in which $\mathfrak{S}_1$ is true; in other words, it is the domain of possibilities left open by $\mathfrak{S}_1$' in the 'object-domain with which S is concerned' (LSL, p. 199). Or, as we would now say: the class of all the possible state-descriptions of the object-domain which make the sentence true. In his *Introduction to Semantics*, Carnap admits that the concept of range is 'primarily a semantic L-concept' and that it should be defined as such (1942, p. 248). In further paragraphs of LSL, Carnap gives characterizations of the 'translation from one language into another' (§61) and of 'the interpretation of a language' (§62), without ever deviating from the basic principles of the syntactical method.

The concept of 'quasi-syntactical sentence' plays a crucial role in Carnap's discussion of the three following issues: extensionality, modalities, and the character of philosophical sentences. Its formal definition is given in §63. Intuitively, a property $E_1$ of an object $c$ is said to be quasi-syntactical when there exists a property $E_2$ applying to *names* of objects and such that $E_2$('$c$') is true exactly when $E_1(c)$ is true. $E_2$ is the syntactical counterpart of the seemingly object-property $E_1$. A *sentence* is said to be quasi-syntactical when it ascribes a quasi-syntactical property to an object (LSL, p. 234). So, the formal definition of quasi-syntactical sentences aims to capture the idea of syntactical sentences in the guise of object-sentences. Carnap argues that the translation of quasi-syntactical sentences into syntactical sentences is possible and he distinguishes several cases where such a translation has an important clarifying effect. In §§65–7, intensional sentences are analysed as a kind of quasi-syntactical sentences which may be translated into (extensional) syntactical sentences. The thesis of extensionality becomes: every intensional language may be translated into an extensional language (LSL, p. 245), and it is stated 'only as a supposition' (LSL, p. 247). Intensional sentences of a specific kind occur in the logic of modalities. In §§69–71, Carnap argues that '*every intensional system of the logic of modalities* [...] *can be translated into an extensional syntactical language*' (LSL, p. 256), so that (in agreement with the basic idea of the syntactical method) '*a special logic of meaning* [*Sinnlogik*] *is superfluous*' (LSL, p. 259). Again, the conclusion is that '*logic is syntax*' (ibid.). The third main application of the concept 'quasi-syntactical sentence' is developed in Part V, where a great number of philosophical sentences are analysed as syntactical sentences in the guise of object-sentences.[69]

## 10 One language?

In §18, Carnap introduces Part II of LSL in the following way:

> Up to the present, we have differentiated between the object-language and the syntax-language in which the syntax of the object-language is formulated. Are these necessarily two separate languages? [...] we intend to show that, actually, it is possible to manage with one language only; not, however, by renouncing syntax, but by demonstrating that without the emergence of any contradictions the syntax of this language can be formulated within this language itself. (LSL, p. 53)

Taken in isolation, this quotation easily leads to misunderstandings and on the face of it the statement that 'it is possible to manage with one language' is surprising in the context of the logical pluralism espoused in LSL. It actually seems to reflect an earlier state of Carnap's thought[70] but what does it mean exactly here?

In the proof of his incompleteness theorem, Gödel had shown that sentences *about* a language $\mathcal{L}$ may be formulated *in* $\mathcal{L}$, provided that $\mathcal{L}$ includes enough arithmetic and is rich enough in means of expression. In Part II, Carnap applies Gödel's technique of arithmetization to Language I: §19 explains how to represent symbols, sentences, and formal proofs by natural numbers (LSL, pp. 55–6), so that syntactical predicates can take the form of arithmetical predicates. In §§20–3, a long series of definitions is provided, which shows how arithmetized syntactical predicates can be defined in Language I itself. However, because of the limited means of expression of Language I, the process of arithmetizing the syntax of I *in* I has a limit:

> The concepts 'derivable' and 'demonstrable' are *indefinite*. [...] If indefinite syntactical concepts are to be defined as well, then an indefinite language must be taken as the syntax-language – such as, for instance, our Language II. (LSL, pp. 75–6)

The non-definability of 'derivable' and 'demonstrable' in the object-language depends on the features of Language I: whereas 'derivable in I' and 'demonstrable in I' are not definable in I, 'derivable in II' and 'demonstrable in II' *are* definable in II. However, a second limitation which does not depend on the choice of the language shows up if we try to formulate in the object-language S itself not only d-terms but also c-terms such as 'analytic in S'. This crucial point,[71] which holds for *any* language, is made in §60c (§60 of the German edition): 'If S is consistent, or, at least, non-contradictory, then *"analytic (in* S)" *is indefinable in* S'[72] (LSL, p. 219). This result is given in the context of an analysis of the antinomies, just after the discussion of 'true' and 'false' in §60b. It is actually a consequence of what is known today as Tarski's theorem on the undefinability of truth (to the effect that 'true in S' is not definable in S if S is consistent).[73]

All this shows that the statement 'it is possible to manage with one language only' needs to be qualified, to say the least. Carnap's point, when he writes this on p. 53, is that it is possible to formulate syntactical sentences in the object-language 'without the emergence of any contradiction', a point which was not at all clear at the time he was working on LSL and thus deserved to be underlined.[74] In June 1931, Carnap gave three lectures in the Vienna Circle. As the minutes taken by Rose Rand testify, at the time he gave these lectures he thought that there was only one language, and that all the sentences, 'even the metalogical ones, are in a *single* language'.[75] By contrast, in LSL, Carnap is perfectly clear about the fact that no single language is sufficient for mathematics: '*mathematics cannot be exhausted by one system*; it requires an infinite series of ever richer languages' (LSL, p. 222).

If this is so, if all the c-terms cannot be defined by means of the syntactical method without having recourse to a richer syntax-language anyway, why bother with arithmetization of some syntactical terms in the first place? The main reason is that the definition of 'provable' and consequently the proof of some crucial syntactical sentences like (Gödel's) incompleteness theorem for Language II are easier – and easier to generalize to all possible language forms – when one can use the arithmetic of natural numbers and the well-known theorems of arithmetic. This is, Carnap explains, '*the most important reason for the arithmetization of syntax*' (LSL, p. 58).[76]

It is essential to distinguish several steps in the process of arithmetization: first, symbols, sentences and proofs are coded by natural numbers in the syntax-language; second, and consequently, the syntactical predicates are given the form of arithmetical predicates of the syntax-language. Formulating these arithmetical predicates in the *object-language* is the third and final step, which cannot always be taken. For example, although 'analytic in S' can be formulated as an arithmetical predicate in the metalanguage (provided this metalanguage is rich enough), this predicate *cannot* be formulated using the arithmetical and logical means of the object-language. So, when we read that pure syntax is 'nothing other than a part of arithmetic' (LSL,

p. 76) and 'nothing more than *combinatorial analysis*, or, in other words, the *geometry* of finite, discrete, serial structures of a particular kind' (LSL, p. 7), we must keep in mind that a c-term of pure syntax like 'analytic in Language II' is an 'indefinite' concept and that its definition as an arithmetical predicate of the metalanguage requires means of expression which are much richer than ordinary arithmetic or what is usually called 'combinatorial analysis'.

Pure syntax, the main subject matter of LSL, abstracts from considerations such as the shape of the symbols, their physical realization, or their occurrence in particular places. These questions are dealt with in *descriptive* syntax, which makes an essential use of descriptive symbols and, for this reason, 'goes beyond the boundaries of arithmetic' (LSL, p. 76).[77]

The question then arises whether pure syntax, which

> is concerned with the possible arrangements, without references either to the nature of the things which constitute the various elements, or to the question as to which of the possible arrangements of these elements are anywhere actually realized (LSL, p. 7)

has anything to do with spoken or written word-languages such as German or French. Regarding this general issue, several more specific questions may be raised. For example: is it possible to apply the method of logical syntax to natural word-languages? How are languages such as I and II meant to be used with respect to word-languages? Are they meant to be auxiliary scientific languages? Or to be imbedded into natural word-languages? Or to replace natural languages in some of their actual uses? Are they supposed to be *coordinated* to ordinary word-languages in some way or other? The reader will not find any extensive discussion of these issues in LSL, but Carnap gives hints at different points. §62 – which deals with the *interpretation* of a language – offers an answer to the first question:

> in the case of an individual language like German, the construction of the syntax of that language means the construction of a calculus which fulfils the condition of being in agreement with the actual historical habits of speech of German-speaking people. And the construction of the calculus must take place entirely within the domain of formal syntax, although the decision as to whether the calculus fulfils the given condition is not a logical but an historical and empirical one, which lies outside the domain of pure syntax. (LSL, p. 228)

So Carnap does not exclude the construction of the logical syntax of natural languages, although he admits that 'the statement of their rules of formation and transformation would be so complicated that it would hardly be feasible in practice' (LSL, p. 2) and although we may wonder what it would mean, in this context, for a calculus to be 'in agreement' with the actual speech habits of people speaking some specific word-language.[78] In §62, Carnap gives hints on how Language I and Language II may be translated or embedded into a word-language, in the same way as a system of geometrical axioms, first given as a calculus, can be either translated or included into the language of physics.[79] He also gives several examples of interpretations of Language II which consist in a translation of II into a word-language (LSL, pp. 230–1) and further argues that a *descriptive* symbol of a language $S_1$ may be translated either as a *descriptive* or as a *logical* symbol of the language $S_2$ into which $S_1$ is interpreted (LSL, pp. 231–3). The consequences of this remark become conspicuous in later writings when Carnap distinguishes between *possible* and *customary* interpretations of arithmetical and geometrical calculi.[80]

## 11 Conventionalism

According to Carnap's principle of tolerance, discussions about the adoption of a form of language should not be constrained by any ideal of correctness. They should rather aim 'to arrive at conventions' (LSL, p. 51). After considering several methods for the treatment of equality, Carnap remarks:

> Philosophical discussions concerning the justification of these various methods seem to us to be wrong. The whole thing is only a question of the establishment of a convention whose technical efficiency can be discussed. (LSL, p. 49)

Similarly, when we set up a language 'it is a matter of convention whether we formulate only L-rules, or include P-rules as well' (LSL, p. 186). The principle of tolerance applies also when we decide whether we want to include indefinite terms in a language:

> Our attitude toward the question of indefinite terms conforms to the principle of tolerance; in constructing a language we can either exclude such terms (as we have done in Language I) or admit them (as in Language II). It is a matter to be decided by convention. (LSL, p. 165)

Finally, about Wittgenstein's 'absolutist conception of language' (in the *Tractatus*), Carnap notes that it 'leaves out the conventional factor in language-construction' (LSL, p. 186).

In view of such quotations, it is no wonder that Carnap's philosophical position in LSL has often been characterized as a form of conventionalism.[81] It has also often been criticized as such. Schlick rejected Carnap's admission of laws of nature (as P-rules) among the defining rules of a language on the grounds that laws of nature are not conventions.[82] Gödel used his incompleteness theorems in order to attempt a refutation of a kind of conventionalism of which Carnap was taken to be one of the upholders (Gödel 1995). As for Quine, he attributed to Carnap a 'linguistic doctrine of logical truth' according to which 'logical truths are true by linguistic convention' (1963, p. 391) and he famously objected that

> the logical truths, being infinite in number, must be given by general conventions rather than singly; and logic is needed then to begin with, in the metatheory, in order to apply the general conventions to individual cases. (ibid.)[83]

So, according to Quine, logical truths cannot be defined conventionally without presupposing logic. Carnap responded to Quine (Carnap 1963b, pp. 915–22) and suggested that formulations such as 'linguistic conventions' be avoided, because he no longer considered them as 'psychologically helpful' (1963b, p. 915). On Carnap's view, choosing L-rules for a language should not be confused with *explaining* what a logical truth *really is*.[84] On the other hand, Carnap did not have a chance to give an answer to Gödel's paper which was only published posthumously in 1995. Today, the significance of Gödel's incompleteness theorems for Carnap's philosophical programme is still one of the most disputed issues in the literature on LSL.[85]

Gödel argues that if the rules defining mathematics are taken to be conventional, 'what must be known is that the rules, by themselves, do not imply the truth or falsehood of any proposition expressing an empirical fact' (Gödel 1995, p. 357). But this comes down to proving that the rules are consistent. By Gödel's second incompleteness theorem, a consistency proof requires mathematical means stronger than the ones to which the proof applies. Therefore, we cannot take mathematical rules to be conventional without presupposing stronger mathematical principles. Gödel wrote no less than six versions of the paper in which this argument is expounded but he never let them be published.[86] Steve Awodey and André Carus have remarked that while Gödel is right in asking that the rules of a language be consistent, he is wrong in asking that they be *demonstrably* consistent (2004, p. 208). Michael Friedman has also objected to Gödel's demand (1999a, pp. 226–7). As for Warren Goldfarb and Thomas Ricketts, they have voiced their own dissatisfaction with Gödel's argument in the following way:

> Gödel's criticism assumes that, on the conventionalist views, we first have a realm of empirical fact; given it, we then adopt the conventions that yield mathematics. In Carnap's terms, this is just to presuppose a language-transcendent notion of empirical fact. But Carnap rejects any such language-transcendent notion. This rejection is part of the message of the Principle of Tolerance. (1992, p. 65)

The distinction between logical and descriptive symbols, between L-rules and P-rules, and between analytic and synthetic sentences of a language can be made only *after* the rules of the language have been conventionally chosen (on pragmatic grounds). In fact, the distinction between the conventional and the factual is determined by the rules of a given language and by the syntactical definitions of 'logical/descriptive', 'L-rules/P-rules', and 'analytic/synthetic' given in §§50–52. This shows that according to the standards of the syntactical method, the distinction between the conventional and the factual is dependent on the rules of the language, not vice versa. No realm of empirical facts is presupposed by the definition of a language.

We may wonder, however, whether this view of the matter gives a sufficient account of the distinction between L-rules and P-rules and, therefore, between 'analytic in S' and 'synthetic in S', a distinction which is of fundamental importance even if relative to the framework of a given language S.[87] Although both kinds of rules may be used to define a language and although even the construction of a physical language is effected '*by means of conventions*' (LSL, p. 320), the P-rules nevertheless embody laws (or even empirical observational sentences) which 'have the character of *hypotheses* in relation to the protocol-sentences' (LSL, p. 318) and which therefore 'can and must be tested by experience' (LSL, p. 320). In §82, Carnap gives a brief account of the procedure by which hypotheses are tested and he asserts the thesis of underdetermination of theory by experience:

> That hypotheses, in spite of their subordination to empirical control by means of the protocol-sentences, nevertheless contain a conventional element is due to the fact that the system of hypotheses is never univocally determined by empirical material, however rich it may be. (LSL, p. 320)[88]

This raises a serious issue regarding the conventional character of P-rules and L-rules: does the word 'convention' mean the same for both kinds of rules? And does a change in the P-rules have to take into consideration the empirical facts, even though P-rules are among the defining rules of a language? How is this compatible with the principle of tolerance?

Although these questions along with other related ones have been at the centre of recent important discussions about Carnap's conventionalism, this is not the place to examine in more detail Carnap's views on this issue and the far-reaching debates in which commentators clarify their own understanding of them. The purpose of the foregoing remarks is only to convey an idea of their philosophical interest and of the importance of providing a precise assessment of the exact consequences of Gödel's incompleteness theorems for Carnap's philosophical programme in LSL. More generally, it is hoped that this introduction and the essays collected in this volume will help the reader make her own way in LSL and make her own decision on how to resolve the interpretive and philosophical issues about which commentators have not yet been able to reach an agreement.

## Gothic letters used for the definition of Language I

| Gothic letters and expressions | Corresponding Latin letters and expressions | First occurrence in LSL | Meaning |
|---|---|---|---|
| $\mathfrak{a}$ | a | p. 17 | symbol (*Zeichen*) |
| $\mathfrak{z}$ | z | p. 17 | numerical variable (*Zahlvariable*) |
| $\mathfrak{nu}$ | nu | p. 17 | symbol '0' (*null*) |
| $\mathfrak{zz}$ | zz | p. 17 | numeral (*Zahlzeichen*) |
| $\mathfrak{pr}$ | pr | p. 17 | predicate (*Prädikat*) |
| $\mathfrak{pr}^n$ | $pr^n$ | p. 17 | *n*-termed predicate (*n-stelliges Prädikat*) |
| $\mathfrak{fu}$ | fu | p. 17 | functor (*Funktor*) |
| $\mathfrak{fu}^n$ | $fu^n$ | p. 17 | *n*-termed functor (*n-stelliger Funktor*) |
| $\mathfrak{verkn}$ | verkn | p. 17 | junction-symbol (*Verknüpfungszeichen*) |
| $\mathfrak{A}$ | A | p. 17 | expression (*Ausdruck*) |
| $\mathfrak{Z}$ | Z | p. 17 | numerical expression (*Zahlausdruck*) |
| $\mathfrak{S}$ | S | p. 17 | sentence (*Satz*) |
| $\mathfrak{A}_{\mathfrak{l}}$ | $A_l$ | p. 25 | logical expression (*logischer Ausdruck*) |
| $\mathfrak{A}_{\mathfrak{d}}$ | $A_d$ | p. 25 | descriptive expression (*descriptiver Ausdruck*) |
| $\mathfrak{St}$ | St | p. 26 | accented expression (*Strichausdruck*) |
| $\mathfrak{Arg}^n$ | $Arg^n$ | p. 26 | argument-expression (*Argumentausdruck*) |
| $\mathfrak{K}$ | K | p. 37 | class of expressions (*Klasse von Ausdrücken*) |

**Further gothic letters introduced for the definition of Language II**

| | | | |
|---|---|---|---|
| $\mathfrak{p}$ | p | p. 84 | predicate-variable (*Prädikatvariable*) |
| $\mathfrak{f}$ | f | p. 84 | functor-variable (*Funktovariable*) |
| $\mathfrak{s}$ | s | p. 84 | sentential-variable (*Satzvariable*) |
| $\mathfrak{v}$ | v | p. 84 | variable (*Variable*) |
| $\mathfrak{k}$ | k | p. 84 | constant (*Konstant*) |
| $\mathfrak{Pr}$ | Pr | p. 83 | predicate-expression (*Prädikatausdruck*) |
| $\mathfrak{Fu}$ | Fu | p. 84 | functor-expression (*Funktorausdruck*) |
| $\mathfrak{sa}$ | sa | p. 84 | sentential symbols (*Satzzeichen*) |
| $\mathfrak{N}$ | N | p. 103 | '0 = 0' |
| $\mathfrak{B}$ | B | p. 107 | valuation (*Bewertung*) |
| $\mathfrak{b}$ | b | p. 108 | valuable symbol (*bewertbares Zeichen*) |

**Further gothic letters introduced for general syntax**

| | | | |
|---|---|---|---|
| $\mathfrak{R}$ | R | p. 186 | series (*Reihe*) |
| $\mathfrak{Stu}$ | Stu | p. 187 | level (*Stufe*) |
| $\mathfrak{Ag}$ | Ag | p. 187 | expressional framework (*Ausdrucksgerüst*) |
| $\mathfrak{Sg}$ | Sg | p. 187 | sentential framework (*Satzgerüst*) |
| $\mathfrak{V}$ | V | p. 191 | variable-expression (*Variabelausdruck*) |
| $\mathfrak{Op}$ | Op | p. 191 | operator (*Operator*) |
| $\mathfrak{Afu}$ | Afu | p. 191 | expressional function (*Ausdruckfunktion*) |
| $\mathfrak{Sfu}$ | Sfu | p. 191 | sentential function (*Satzfunktion*) |
| $\mathfrak{W}_1$ | $W_1$ | p. 199 | range (*Spielraum*) |
| $\mathfrak{Vk}$ | Vk | p. 201 | sentential junction (*Satzverknüpfung*) |
| $\mathfrak{vk}$ | vk | p. 201 | junction-symbol (*Verknüpfungszeichen*) |
| $\mathfrak{zpr}$ | zpr | p. 205 | numerical predicate (*Zahlprädikat*) |
| $\mathfrak{zfu}$ | zfu | p. 205 | numerical functor (*Zahlfunktor*) |
| $\mathfrak{Q}$ | Q | p. 222 | syntactical correlation (*syntaktische Zuordnung*) |

## Other gothic letters used in LSL

| | | | |
|---|---|---|---|
| $\mathfrak{G}$ | G | p. 130 | the Gödel sentence |
| $\mathfrak{W}_{II}$ | $W_{II}$ | p. 133 | some particular sentence defined on p. 133 |
| $\mathfrak{W}$ | W | p. 214 | true (*wahr*) |
| $\mathfrak{F}$ | F | p. 214 | false (*falsch*) |
| $\mathfrak{N}$ | N | p. 214 | non-sentence (*Nicht-Satz*) |

## Notes

1. These lectures have been published in Creath (1990, pp. 47–103).
2. Some examples of authors holding it are given in Carus (1999, p. 16), and (2007, pp. 32–7). For a critique of such attitude, see also Creath (1991).
3. This series, entitled 'Schriften zur wissenschaftlichen Weltauffassung' (Writings for the scientific conception of the world) includes Carnap's *Abriss der Logistik* (*Outline of Logic*) (1929), and Popper's *Logik der Forschung* (1935) (later translated as *The Logic of Scientific Discovery*).
4. The translation was much revised by several people, including Olaf Helmer, before its publication.
5. The content of §§34a–i was published as Carnap (1935d), and the content of §§60a–d and §§71a–d as Carnap (1934e), with some modifications.
6. The German original of this paper, which can be found in Carnap (2004a), had not been published before.
7. Carnap gave two other papers at this congress: Carnap (1936b) and (1936c). The second one already reflects the evolution of Carnap's thought toward the adoption of semantic concepts.
8. In 1934, an English translation of this paper was published as a book under the title *The Unity of Science* (1934b), with two introductions (one by Max Black and one by Carnap) and with corrections by Carnap.
9. The principle of tolerance first appeared in print in Carnap (1932c).
10. The term 'logical syntax', which had been used by Wittgenstein in the *Tractatus*, appears in Carnap (1932a, p. 228).
11. See Awodey and Carus (2007) and their contribution to this volume, as well as Uebel (2005), Uebel (2007a, pp. 140–50), Carus (2007), and Richard Creath's contribution to this volume.
12. We occasionally indicate the original German word in parentheses, when this clarifies matters, or when the German word is frequently used by commentators. Issues of translation were discussed in the Carnap–Quine correspondence while the translation of *Logische Syntax der Sprache* was in progress (see Creath 1990, p. 132 and pp. 136–44). Other terminological remarks about LSL are to be found in Carnap (1942, §39), a book that Carnap wrote in English. But the translation issues discussed there are complicated by the evolution of Carnap's philosophical viewpoint, from syntax to semantics, on which Carnap also comments in the same paragraph.

13. This phrase does not occur in major writings such as Carnap (1936–37), Carnap (1939), and Carnap (1942) (except in the appendix, §39, where the differences with the views of LSL are discussed). It does occur in Carnap (1938), where the author writes: 'we may distinguish between logic of science in the narrow sense, as the syntax of the language of science, and logic in the wider sense, comprehending both syntax and semantics' (pp. 44–5).
14. The title of §73 reads: 'The logic of science is the syntax of the language of science'.
15. On LSL in the context of the Vienna Circle, see Thomas Uebel's contribution to this volume, and Uebel (2005), (2007a), (2007b).
16. In particular, it is not Carnap's intention to leave aside philosophical problems and concentrate on scientific ones. See Friedman (1999a/1999b, pp. 210–14). On Carnap's philosophic programme in LSL, see Richard Creath's contribution to this volume.
17. 'My endeavour in these pages is to explain the main features of the *method of philosophising* which we, the Vienna Circle, use, and, by using try to develop further' (1935a, p. 6). On the other hand, Carnap also writes: 'We [the members of the Vienna Circle] give no answer to philosophical questions, and instead *reject all philosophical questions*, whether of Metaphysics, Ethics or Epistemology. For our concern is with *Logical Analysis*. If that pursuit is still to be called Philosophy, let it be so; but it involves excluding from consideration all the traditional problems of Philosophy' (1934b, pp. 21–2). Here and everywhere else in this introduction, emphases in quotations are from the original.
18. See LSL, p. 299.
19. The specific features of Language I and Language II which make this completeness result possible and the link between determinacy and completeness will be examined below.
20. See Carnap (1929), (1930a), (1930b), and (1931a).
21. Church (1932) is one of the seldom noticed possible sources of Carnap's principle of tolerance: 'We do not attach any character of uniqueness or absolute truth to any particular system of logic. The entities of formal logic are abstractions, invented because of their use in describing and systematizing facts of experience or observation, and their properties, determined in rough outline by this intended use, depend for their exact character on the arbitrary choice of the inventor. [...] There exist, undoubtedly, more than one formal system whose use as a logic is feasible, and of these systems one may be more pleasing or more convenient than another, but it cannot be said that one is right and the other wrong', Church (1932, pp. 348–9). This paper was published in April 1932, a few months before Carnap's adoption of the principle of tolerance, probably in October 1932 (see Carus 2007, p. 252 and Awodey and Carus, *infra*, p. 97) and it is mentioned in the bibliography of the 1934 German edition of LSL. This is not to say that there is nothing more in Carnap's principle of tolerance than in Church (1932).
22. There was a time, however, when the principle of tolerance was ignored by commentators. See Carus (2007, p. 35). The principle of tolerance also has a large place among contemporary philosophers of logic who try to provide an account of logical pluralism from a technical viewpoint.
23. In LSL and in his 'Intellectual Autobiography', Carnap acknowledges the deep influence these authors had on his philosophical thinking. On these influences and on the ways Carnap nonetheless dissents from their views, see for example Ricketts (1994), Friedman (1997), Carus (2007, chs. 5 and 7).

24. This point is made forcefully by Ricketts (1994). See also Ricketts (2007).
25. This would require some qualification for some parts of mathematics such as geometry. See LSL, §71e.
26. The issue of the relationships between natural word-languages and calculi in LSL is controversial. See Ricketts (2003), (2004), Carus (2007, pp. 245–50 and pp. 273–84), and Thomas Ricketts's contribution to this volume. See also below, pp. 39–40.
27. The syntactical definition of 'definite' (*definit*) and 'undefinite' (*indefinit*) is given in §15 (LSL, pp. 45–6). The German word '*definit*' could have been rendered in English by 'finitary'. In §34i, Carnap remarks: 'What is meant by "finite means" is not stated exactly in any work of Hilbert which has been published up to now [...], but presumably what we call "definite syntactical concepts" is intended' (LSL, pp. 128–9).
28. Here 'S' is used as an abbreviation of '*Sprache*' (language). Throughout LSL, in the English translation as well as in the German original, 'S', '$S_1$', '$S_2$', etc. are used as names for languages, not 'L', '$L_1$', '$L_2$', etc.
29. Gödel's argument and some objections to it are examined below, p. 41.
30. A list of the gothic symbols used LSL together with their meaning and the equivalent in the Latin alphabet is provided below, pp. 43–5.
31. The reader who is familiar with contemporary deductive logic can skip this section. On the other hand, the interested reader can study, for example, Enderton (1972/2001) but what follows is not a survey of this particular book.
32. This qualification will become clear below.
33. The language $\mathcal{L}$ we are now considering is said to be 'first-order' because any variable of $\mathcal{L}$ ranges over individuals of $\mathcal{D}$ and not, for instance, over subsets of $\mathcal{D}$. In such a language, no expression can be interpreted as quantification over *properties* of these individuals.
34. 'Material' (*inhaltlich*) is meant here as the English correlative of 'formal'. See the Carnap–Quine correspondence in Creath (1990, p. 132 and p. 136).
35. Translation modified.
36. Though the formulation of axiom systems is possible. It is discussed in §71e.
37. See above, p. 46, n. 20.
38. See LSL, p. 327, Goldfarb and Ricketts (1992), and Warren Goldfarb's contribution to this volume.
39. On Carnap's concept of logical consequence in the pre-syntactical period, see Carnap (2000), Awodey and Carus (2001), Reck (2007), and the references given in these papers.
40. These rules are modus ponens and an informal version of the rule of substitution. See Carnap (1929, p. 11).
41. On this issue, see Philippe de Rouilhan's contribution to this volume.
42. The German translation of Tarski's paper on the concept of truth was first published in 1935, the Polish original in 1933. Carnap had heard about it from Gödel who had arrived at the basic idea independently in 1931–2, but Carnap thought of it as only a definition of logical or mathematical truth. See Awodey and Carus (2007, p. 37), and the quotation below, p. 28.
43. Throughout LSL, '*entscheidbar/unentscheidbar*' is translated by 'resoluble/irresoluble'. This translation was suggested by Quine (see Creath 1990, p. 142) but is not in use any more. In his Princeton Lectures (1934), Gödel already used the expression 'undecidable propositions'.
44. In LSL, when Carnap writes 'rules of inference' (*Schlussregeln*), what he has in mind are *finite* rules. Only 'rules of consequence' (*Folgebestimmungen*) may be

infinite. The generic term for rules that are either finite or infinite is 'rules of transformation' (*Umformungsbestimmungen*).

45. A system of formal derivation (a 'deductive calculus') for first-order logic is given, for example, in Enderton (1972/2001, pp. 109ff.).
46. For a definition of 'logical consequence' (some authors say 'logical implication') see Enderton (1972/2001, p. 88).
47. For a comparison of Carnap's approach and Tarski's, see Coffa (1987, pp. 551ff.), Oberdan (1992, pp. 254–5), and Philippe de Rouilhan's contribution to this volume.
48. '$\mathfrak{S}_1$' designates some particular sentence of the object-language S. Gothic letters with subscript integers designate particular expressions; other gothic letters usually designate categories of expressions, though there are exceptions: e.g. $\mathfrak{G}$ (to be read 'G') designates a particular sentence (LSL, p. 130).
49. Coffa argues that 'truth *can* be defined in what [Carnap] calls the syntax of a language'. According to Coffa, 'this is obvious as soon as we realize that Carnap allowed his syntax to include translations of their object-languages' (Coffa 1987, p. 567). This argument is mistaken. The languages Carnap takes as syntax-languages (e.g. English with the addition of gothic symbols) have resources that go far beyond the limits of the syntactical method anyway. The point is that, except in special cases (e.g. object-languages with a finite number of synthetic sentences), a Tarski-like definition of truth based on the translation of the expressions of the object-language in the metalanguage requires more than a reference to the kinds and order of the symbols from which these expressions are constructed, i.e. more than what Carnap's method of logical syntax can offer.
50. The first extensive discussion of Carnap's views on truth in LSL is Kokoszyńska (1936).
51. In 1922, Carnap already wrote 'We begin the construction of formal logic with the undefined basic concepts "true" and "false"' (1922, p. 9).
52. In a previous stage of his project, Carnap started out with *one* canonical language, and this was what is known as 'Language I' in LSL.
53. See for instance Menger's paper 'On Intuitionism' (1930).
54. '$\mathfrak{Z}_1$' (the gothic writing for '$Z_1$') designates a numerical expression: it may be a numerical variable ($\mathfrak{z}$), a constant numeral ('0', '$0^{|}$'...), or an expression involving function-symbols or the K-operator.
55. '*Rekursive Definition*' is translated as 'regressive definition' in LSL.
56. Translation modified.
57. There is a link between this kind of transformation and Carnap's ideal of explication. See Beaney (2004) and Carus (2007, ch. 11).
58. See, among others, Tarski (1936b), Mac Lane (1938), Kleene (1939a), Coffa (1987), Oberdan (1992), Sarkar (1992), Awodey and Carus (2007, pp. 35–8), and Philippe de Rouilhan (this volume).
59. On this crucial issue, see Friedman (1988/1999b, p. 172), Awodey and Carus (2007, pp. 37–8), and Warren Goldfarb's contribution to this volume.
60. '$\mathfrak{A}_1$' (to be read '$A_1$') designates an expression (*Ausdruck*) of the object-language.
61. '$\mathfrak{K}_1$' (to be read '$K_1$') designates a class (*Klasse*) of expressions of the object-language.
62. The word 'valid', like any other syntactical term, should be supplemented by 'in S' (for some S of which we know only that it is characterized by some (unknown) formation and transformation rules). Following Carnap, we usually leave this addition implicit.

63. In fact, the definition of 'logical' given in §50 is known to be defective. An analysis of this definition, its defects, and proposed amendments can be found in Creath (1996), Creath (forthcoming), and Denis Bonnay's contribution to this volume.
64. See also LSL, p. 186 and pp. 316–22.
65. See Schlick (1936), Uebel (2007a, pp. 342–56), and Thomas Uebel's contribution to this volume.
66. On this issue, see Coffa (1991), Friedman (1988), Friedman (1994/1999b, pp. 68–70), Friedman (1999a), Oberdan (2004). It may be added that those who did accept the notion of P-rules often seriously misunderstood Carnap's proposal. The example of Wilfrid Sellars is examined in Carus (2004).
67. On the evolution of this debate, which started only much later, a few years after the Second World War, see Creath (1990, Introduction).
68. See Ricketts (1994, pp. 189–93) and Friedman (1999a).
69. See Carus (2007, pp. 256–61), and Jacques Bouveresse's and Pierre Wagner's contributions to this volume.
70. Carnap had hoped to be able to manage with one language until January 1931, and then again from about June 1931 to October 1932. Carnap (1932b), written in late 1931, testifies to this hope. On the evolution of Carnap's thought on this point in 1931–2, see Carus (2007, ch. 9) and Uebel (2007a, pp. 140–50).
71. Its discovery in 1931 had a deep impact on Carnap's philosophical programme. See Awodey and Carus (2007, pp. 35–8) and Carus (2007, p. 251).
72. A language S is *non-contradictory* if at least one sentence is not *demonstrable*; *consistent* if at least one sentence is not *valid*. See §59 (LSL, p. 207).
73. On the definability of 'analytic in II', see also LSL, p. 113, as well as Oberdan (1992, pp. 255–6) and Ricketts (1996, p. 234).
74. Carnap quotes Wittgenstein and Herbrand as examples of authors having different views. For a comparison of Carnap and Wittgenstein in this context, see Friedman (1997).
75. Stadler (1997/2001, p. 329), quoted by Awodey and Carus, below, p. 105, n. 15.
76. See Carnap's own comments on his adoption of an arithmetized metalogic, quoted in Stadler (1997, p. 325).
77. On descriptive syntax, see LSL, p. 7, pp. 53–4, and §§24–5.
78. See also the different issue raised in Ricketts (2003, p. 261). On the relationships between calculi and word-languages, see the references given above, p. 47, n. 26, especially Thomas Ricketts's contribution to this volume.
79. This example is given in LSL, p. 229.
80. See Carnap (1939a, pp. 39–56) and Friedman (2008).
81. On Carnap's conventionalism, see Creath (1992).
82. See Schlick (1936) and above, p. 35.
83. This argument had already been expounded in Quine (1936) although it was not formulated there as an objection to Carnap (see Creath 1990, p. 30).
84. See Goldfarb and Ricketts (1992, p. 71). See also Richardson (1997).
85. See Friedman (1988), (1999a), Goldfarb and Ricketts (1992), Ricketts (1994), (1996), Richardson (1994), Goldfarb (1995), Potter (2000, ch. 11), Awodey and Carus (2003), (2004), and the other references given in Awodey and Carus (2004, p. 203, n. 1).
86. Version III and version V have been published as Gödel (1995).
87. This point is made by Michael Friedman (1999a, pp. 215–20).
88. Carnap also asserts a form of holism known today as the Duhem thesis: '*the test applies, at bottom, not to a single hypothesis but to the whole system of hypotheses* (Duhem, Poincaré)' (LSL, p. 318).

# Part I

# The Route to *The Logical Syntax of Language*

# 1
# Carnap's *Logical Syntax* in the Context of the Vienna Circle

*Thomas Uebel*

Rudolf Carnap's *Logical Syntax of Language* constitutes one of the towering achievements of the second generation of the philosophers of the 'new' logic, not far behind Kurt Gödel's work on the incompleteness of arithmetic or Alfred Tarski's on the semantic conception of truth. But the publication of *Logical Syntax* in 1934 was a significant event not only for philosophy generally – though widely recognized as such only much later – but also for the Vienna Circle. Moreover, it was clearly understood to be such an event. In this chapter, *Logical Syntax* will be considered not so much in systematic but in historical terms, in the context of a discursive field with numerous voices, the Vienna Circle and its collaborators. Even thus delimited, not all of Carnap's interlocutors can be considered and Gödel and Tarski, whose role is discussed in other chapters, must be disregarded. Instead the focus lies on how Carnap's *Logical Syntax* fits into the dynamic of the overall development of the Vienna Circle.

The thesis argued for is that Carnap's *Logical Syntax* not only emerged against a very rich philosophical-mathematical background, but also hastened the Circle's demise as a coherent set of philosophies by further sharpening the divergence of views that had been developing all along. The rich mathematical background is delimited by the work of Karl Menger and Hans Hahn; the intra-Circle consequences of *Logical Syntax* are fittingly illustrated by the testimonies of Moritz Schlick and Otto Neurath. (Needless to say, the ghost of Wittgenstein hovers over the proceedings and will occasionally have its toes trodden on.)

I begin by setting the principal insight and methodological core of *Logical Syntax*, the Principle of Tolerance, against the anticipatory comments about a solution to the foundational debate in mathematics in the Circle's manifesto of 1929 and by considering two claims that cast doubt on the originality of Carnap's conception. Turning then to the reception proper of Carnap's book, Schlick's and Neurath's reactions will be discussed to show that while it was celebrated on the so-called left wing of the Vienna Circle, it was criticized on the so-called right wing. In the final part it will be argued that neither

Schlick's nor the left wing's reactions should be thought surprising: Schlick's indifference to Tarski's metamathematics contrasts tellingly with Hahn's so far insufficiently appreciated efforts in the direction of logical pluralism.

## 1 From the manifesto to *Logical Syntax*

To see the pivotal role that *Logical Syntax* played in the history of the Vienna Circle, we need only recall the Circle's manifesto 'Wissenschaftliche Wetltauffassung. Der Wiener Kreis' of 1929. Already there the foundational dispute in mathematics between logicism, formalism, and intuitionism was given a prominent place in the catalogue of problems of concern. No solution was indicated beyond the vaguely promissory pronouncement:

> Some hold that the three views are not so far apart as it seems. They surmise that essential features of all three will come closer in the course of future development and probably, using the far-reaching ideas of Wittgenstein, will be united in the ultimate solution. The conception of mathematics as tautological in character, which is based on the investigations of Russell and Wittgenstein, is also held by the Vienna Circle. (Carnap, Hahn, Neurath 1929/1973, p. 311)

Five years later, Rudolf Carnap, a co-signer of the manifesto, published *Logical Syntax* in which logicism, formalism, and intuitionism were not so much reconciled as meant to be transcended in a far-reaching reconceptualization of both logic and philosophy. Likewise, the reconstruction of mathematics Carnap developed was 'based on' Russell and Wittgenstein only in a genetic sense: in dogmatic terms it contradicted both in significant respects.

Some of the ideas put forward in *Logical Syntax* were soon abandoned by their author, for instance, the fake syntacticism that abjured semantics and dismissed all talk of meanings. Others are still discussed seriously today, the most prominent of these, and the one to contradict Russell and Wittgenstein (at least as-then-published) most spectacularly, being the 'principle of logical tolerance':

> *In logic there are no morals*. Everyone is at liberty to build his own logic, i.e. his own form of language, as he wishes. All that is required of him is that, if he wishes to discuss it, he must state his methods clearly, and give syntactical rules instead of philosophical arguments. (1934d/1937, §17)[1]

In one exposition, Carnap illustrated the intended dissolution of the foundational dispute with the following example. Take the finitism of Hilbert's programme and consider the claim, and its denial, that in laying the foundations of mathematics no recourse must be had to unrestricted existential propositions:

> If the theses are not meant as assertions but as proposals, the dispute vanishes: one sets up two different languages. [...] If the question is displaced onto the level of syntax, one is [...] concerned [...] with the consequences of the introduction or elimination of these concepts, and especially, with securing freedom from contradiction in their employment. (1934c/1987, p. 65)

The principle of logical tolerance was meant to reorient philosophical work away from the search for metaphysical foundations of truth claims for logic and mathematics to the elaboration and consideration of possible logico-linguistic frameworks that would render explicit the validity of the claims made on conventional grounds. In just this respect, *Logical Syntax* did provide something approaching 'an ultimate solution' to the foundational dispute in mathematics, namely, by demonstrating how it may be argued that there is no fact of the matter as to whether, say, classical or intuitionist arithmetic is 'correct'. Of course, that amounted to a *dissolution* of the problematic itself. What consideration of the Vienna Circle context of Carnap's 1934 work underscores is that the nature of this dissolution of the traditional philosophical problematic was not always correctly perceived, even by other members and associates of the Vienna Circle itself.

## 2 Menger's practical pluralism

Already the quote from the manifesto suggests that the deep philosophical background for Carnap's *Logical Syntax* was Wittgenstein's philosophy of logic in the *Tractatus* (also explored in another chapter). But an important mathematical background in and around the Circle must not be discounted. To begin to see this it will be helpful to consider the reaction to *Logical Syntax* by Karl Menger. Often considered in terms of a priority dispute, indeed first cast as such by Menger himself, it is best understood as illustrating the rich background that inevitably informed Carnap as he developed the project he called 'metalogic', even 'semantics', before settling on 'logical syntax'. Carnap concluded the section which introduced logical tolerance with the following paragraph:

> The tolerant attitude here suggested is, as far as special mathematical calculi are concerned, the attitude which is tacitly shared by the majority of mathematicians. In the conflict over the logical foundations of mathematics, this attitude was represented with especial emphasis (and

> apparently before anyone else) by Menger ['Der Intuitionismus' (1930)]. Menger points out that the concept of constructivity, which intuitionism absolutizes, can be interpreted both in a much narrower and in a much wider sense. The importance for the clarification of the pseudo-problems of philosophy of applying the attitude of tolerance to the form of language as a whole will become clear later [...] (1934d/1937, §17)

This paragraph was the result of an intervention by Menger. In letters to Carnap, Menger had raised his priority claim – he wrote of 'the principle of syntactical tolerance, which I regard as my intellectual property' – already prior to the publication of *Syntax*.[2] Afterwards Menger also objected to Carnap's published acknowledgement.[3] He insisted that he too was concerned with philosophical problems (1933 [in English translation 1937, p. 335 only]; cf. 1979a, p. 16, n.8), but this Carnap had not denied. Much later still, Menger not only repeated that in developing certain still earlier ideas about the variety of possible definitions of 'constructivity' in his 'Bemerkungen zu Grundlagenfragen der Mathematik' (1928), he had hit upon the leading ideas of logical tolerance in 'Der Intuitionismus' (1930) and in his 1932 lecture 'Die neue Logik' (1933), but he also stressed that he did so at a time in the late 1920s when, he claimed, the other members of the Circle still showed no interest in these ideas (1979a, p. 12).

Menger's priority claim has found some acceptance in the literature (eg. Gillies 1981; Köhler 1991), but Michael Friedman has recently argued against it. All that Menger did in 'Intuitionismus' was to represent 'the attitude of the "ordinary working mathematician" to which Carnap allude[d]' in his acknowledgement. That was an attitude of 'stark dismissal' of the foundational debate. 'But the whole point of Carnap's principle of tolerance is to articulate a systematic mode for resolving or dissolving such philosophical disputes' (Friedman 1999a/1999b, p. 211). To assess the matter, let's consider Menger's pronouncements.

The relevant remark in his 'Bemerkungen' is short enough: 'I wish to emphasise, however, that I consider constructivity as a term that has not as yet been made precise and that can be made precise, if at all, then probably *in various ways* and *in various degrees*' (1928/1979b, p. 87 n.9). This remark is repeated and slightly expanded, in his 'Intuitionismus':

> The author has repeatedly expressed the opinion that the heretofore undefined concept of constructivity could be made precise in various ways and degrees. [...] For example, it is perhaps possible to give a constructivity principle so strict that it would allow only finite sets, or a somewhat weaker one which would include countable sets, or a weaker one still which would admit analytic sets, or a very general one which

> would allow arbitrary sets of real numbers. The requirement of consistency may in this sense be considered the weakest possible constructivity principle. (1930/1979b, pp. 56–7)

Against the dogmatic attachment to any one such principle, Menger set his 'implicationist' standpoint: 'what matters in mathematics and logic is not which axioms and rules of inference are chosen, but rather what is derived from them' (ibid.). This is repeated as one of the two relevant central points in 'Die neue Logik':

> *What interests the mathematician and all that he does is to derive propositions by methods which can be chosen in various ways but must be listed, from initial propositions which can be chosen in various ways but must be listed.* And to my mind all that mathematics and logic can say about that activity of the mathematicians (which neither needs justification [*Begründung*] nor can be justified) lies in this simple statement of fact. (1933/1979b, pp. 40–1)

Menger, too, put forward a philosophical position. To be sure, from 1928 he opposed intuitionism on the basis of possible mathematical practice – which, however, he stressed, was overlooked by 'prominent mathematicians' like Poincaré, Hilbert, Weyl, and Brouwer (1979a, p. 13) – but with the arguably philosophical argument that no good reasons for dogmatic restrictions of such practice had been produced. By 1932, moreover, Menger articulated a still stronger philosophical thesis which claims, albeit without much argument, the irrelevance and impossibility of foundations. At the same time he rejected all attempts that, after Gödel, sought to restore the unconditional confidence in mathematical knowledge that had been undermined by the discovery of the set-theoretic paradoxes: such foundations were now forever out of reach.

In that Menger's attention to the foundational dispute extended no further, his position is different from that taken by Carnap, even though Carnap too wrote that in the construction of the language of mathematics 'no question of justification [*Berechtigung*] arises at all, but only the question of the syntactical consequences to which one or another of the choices leads, including the question of non-contradiction'. Yet Carnap did not dismiss attempts to reconstruct the language of mathematics in accordance with various desiderata, but only the 'striving after "correctness"' (1934d/1937, p. xv). Thus he devised two languages of different strength for the reconstruction of (different parts of) arithmetic – Language I fulfilling 'certain conditions laid down by intuitionism' and Language II fulfilling 'simultaneously, the demands of both formalism and logicism' (ibid., §§16, 84) – and he noted:

> When we construct our Language I in such a way that it is a definite [decidable] language [...] we do not mean thereby to suggest that this is the only possible or justifiable form of language. We shall, on the contrary, include the definite Language I as a sub-language in the more comprehensive Language II, and the form of both languages will be looked upon as a matter of convention. (ibid., §16)

For Carnap, the issue of justification appeared in a different light than for Menger and as such it was not dismissed but relativized, with different standards of language choice answering to different pragmatic concerns. Moreover, Carnap's own concern with the foundational dispute centred on the issue of applicability, which Menger showed no interest in at all.[4] So Carnap's 'more philosophical' take on logical tolerance consisted in this: whereas Menger dismissed attempts at justification outright, Carnap relativized them and focused them on the issue of the applicability of mathematics. Menger's priority claim must be rejected on the grounds that his version of logical pluralism differed from Carnap's.

Importantly, however, that his priority claim fails does not mean the denial of the importance of Menger's role in bringing some form of logical pluralism into the Circle's discussions, for it underlines the rich background against which Carnap worked.[5] Indeed, the fact that, in practice, mathematicians were able to show much flexibility in how central concepts could be interpreted or rendered operational will have been perceived as a very significant one by Carnap once he came to reflect upon it. Moreover, Menger was surely right to stress that Carnap did not embrace logical pluralism from the start of his metalogical project: he noted that 'Carnap's belief in uniqueness remained unshakable up to the time of his first visits in Vienna after he had left for Prague in 1931' (1979a, p. 13).[6] But this too does not mean that when the pluralist penny dropped, it did not fall differently for both – even if a further background factor common to both Menger and Carnap must be noted.

When Carnap considered logical pluralism seriously after he had moved to Prague, he did so in the context of reflecting on the potential of the metamathematical approach that Tarski had impressed upon him already during his first visit to Vienna in February 1930.[7] Tarski had been invited to Vienna by Menger who felt that much was to be gained by closer cooperation between the Warsaw logicians and the Vienna Circle. Menger, like Carnap, was disappointed by the indifference to Tarski's work shown by Schlick.[8] That said, however, Carnap looked at logical pluralism in the metamathematical perspective with a particular *Problemstellung* in mind not shared with Menger. Menger's dismissiveness of the foundational dispute in mathematics contrasts with Carnap's attempt to overcome it while answering what he perceived to be the core concern of the philosophy of mathematics, one that was addressed by logicism alone: the applicability of mathematics.

## 3 Schlick's Wittgensteinian intervention

Turning to Schlick, we must note first what borders on a second priority dispute that Carnap had to weather while *Logical Syntax* was in press. Unlike the one just considered which helps to deepen our appreciation of the sophisticated mathematical background against which *Logical Syntax* was conceived, this dispute tells us about the way things were going in the Vienna Circle as we enter the mid-1930s. What we must recall here is that, philosophically, for Carnap, the *Tractatus* provided both a starting point – his conception of logical truths as tautologies – and a conception to be overcome. Michael Friedman articulated a rare consensus amongst interpreters when he wrote that it was Carnap's ambition 'to represent the most general possible logico-mathematical pluralism', with the key move being 'the rejection of the logical absolutism of the *Tractatus*' (1999a, p. 199 and 1997/1999b, p. 183). Carnap must have found particularly baffling, therefore, certain claims repeatedly made by Schlick.

Perhaps mindful of the priority dispute over the inception of physicalism that played itself out in the summer of 1932 in an uncomfortable correspondence between Wittgenstein and Carnap via himself as the middle man – and hoping to avoid a repeat – Schlick informed Carnap in the spring of 1934 of what he perceived as a convergence in his and Wittgenstein's views.[9] Equally mindful, Carnap had composed a note for the Foreword to *Logical Syntax*. After seeing the proofs, Schlick wrote to Carnap again on 2 June 1934. Claiming that 'Wittgenstein has long been convinced of the possibility of an absolutely free choice of linguistic rules' (ASP, RC 029-28-13), Schlick urged Carnap to change his wording of the second sentence in which Schlick's previous report of Wittgenstein's views was reported. The passage read as follows:

> A propos of the remarks made – especially in §17 and §67 – in opposition to Wittgenstein's former dogmatic standpoint, Professor Schlick now informs me that for some years, in writings as yet unpublished, Wittgenstein has taken the view that the rules of language can be chosen freely. Perhaps his view too is developing in the direction of the Principle of Tolerance. (Copy of p. vii of typescript of *Logische Syntax*, ASP, RC 029-28-12)

Schlick objected to its apparent ascription of uncertainty to the reported change of Wittgenstein's mind: he was sure of what he reported. Carnap responded on 4 June that

> I myself do not have the impression that Wittgenstein adopts the conception which I designate as the Principle of Tolerance. To be sure, it seems as if he now adopts a more tolerant conception than he (and we all) adopted earlier on. But according to what I have learnt from you (especially from the last paper) and from Waismann, his views do not coincide wholly with mine on this point. (E.g., he rejects, if I am informed correctly, sentences that cannot be conclusively verified; moreover, you, and so I suspect he as well, allow as analytic sentences (tautologies) only those for which we possess a decision procedure.) We can talk about these questions later on at our leisure. Here what matters is only that I do not believe that we are in agreement. (ASP, RC 029-28-11)

In his response of 5 June, Schlick disputed Carnap's interpretation and claimed that 'since at least four years' Wittgenstein did not hold the view concerning not-conclusively-verifiable statements that Carnap ascribed to him but accepted them as hypotheses, just as he would not formulate his views of tautologies in the fashion in which Carnap reported them: 'his views and ways of expressing himself are in every respect much less restrictive' (ASP, RC 029-28-10). After some discussion Schlick suggested dropping the offending second sentence and adding 'völlig frei' (with complete freedom) in the first sentence.

In his letter of 29 June, Carnap agreed to drop the offending phrase, leaving the published Foreword to read as follows.

> Incidentally, a propos of the remarks made – especially in §17 and §67 – in opposition to Witgenstein's former dogmatic standpoint, Professor Schlick now informs me that for some time past, in writings as yet unpublished, Wittgenstein has agreed that the rules language may be chosen with complete freedom. (Carnap 1937, p. xvi)[10]

Still, in his response Carnap also noted reservations concerning Schlick's interpretation of Wittgenstein's older views.

> But I am not sure whether your interpretation of his views from the time that lies back a few years is correct, nor Waismann's interpretion. I cannot see how these interpretations can be brought into agreement with your reports at the time and especially with the very clear wording of Waismann's Theses of 1930. (ASP, RC 029-28-08)

Clearly, Carnap was not convinced by Schlick's retrospective claims. What Carnap called Waismann's 'Theses of 1930' is a typescript that circulated in and was discussed by the Circle in the years 1930 and 1931.[11] It reads like an updated and clarified interpretation of the *Tractatus* – by means, e.g., of the addition of the verification principle and the concept of hypotheses – and was understood by the members of the Circle to be the fruit of Waismann's talks with its author and a representation of his latest views. In particular, the logical absolutism of the *Tractatus* reappears in the form of conclusively verifiable elementary propositions.[12]

Having nothing but Schlick's interpretive claims about Wittgenstein's more recent ideas to go on, Carnap was understandably unwilling to concede agreement with Wittgenstein on what he took to be the distinctive characteristic of *Logical Syntax*: its overcoming of logical absolutism and the introduction of logical pluralism. Schlick's intervention here foreshadowed his tendency either to discount the distinctiveness of Carnap's theses in *Logical Syntax* or to disagree with them.

## 4 Schlick's criticism of *Logical Syntax*

With the publication of *Logical Syntax*, quite radical differences between Schlick's and Carnap's philosophical orientations began to become apparent.

The first difference between Schlick and Carnap to be noted springs from the fact that Schlick rejected the idea that Carnap gave prominent expression to in *Logical Syntax*, namely, that no statements can ever be conclusively verified. As is clearly stated in 'Sur les constatations' (1935b), Schlick held on to the view expressed in his 'Foundation of Knowledge' (1934) that there exist statements or propositions ('Aussagen'), namely his 'Konstatierungen', that can be and are by their very nature conclusively verified. By contrast, *Logical Syntax* claims that 'not only laws, however, but also concrete sentences are formulated as hypotheses' and that 'there is in the strict sense no refutation (falsification) of an hypothesis [...] still less is there in the strict sense a complete confirmation (verification of an hypothesis)' (Carnap 1937, §82, p. 318).

On 14 November 1934 Schlick wrote:

> The principle of logical tolerance is very nice; in its application one must be very careful, however, that one does not make determinations which earlier conventions already have ruled out or which supercede commonly accepted ones without drawing the reader's attention to it. I believe this is involved in the misunderstandings of the controversy about protocol sentences. For instance, I can very well determine that the sentences 'here now red' and 'I see red' have the same meaning, but with this I do violence to common usage, for in the second sentence there appear in the

> common usage of the words the mention of a human body (I) and of a physiological process (seeing), of which there is absolutely no mention in the first sentence. (ASP, RC 102-70-11)

To be sure, Schlick does not here declare illegitimate Carnap's fallibilist conception of the scientific language, but he claimed that that cannot capture 'the psychological importance of the possibility of another convention' (ibid.), namely that of infallible evidence statements. Schlick reined in the class of possible logico-linguistic frameworks that he deemed of relevance for epistemological analysis. When Carnap 'abandoned' epistemology for the logic of science around this time – see his Paris Congress lecture 'Von der Erkenntnistheorie zur Wissenschaftslogik' (1936a) – he left behind, of course, precisely any such subject-oriented approach to the justification of knowledge claims, including Schlick's.

Admittedly, this was only a divergence in their philosophical interests, not yet an outright conflict. Even so, Carnap's 'confession' in his letter of 4 December 1935, that 'even from your letter and the additional remarks in the French pamphlet, which I knew already, your view of the "Konstatierungen" has still not become clear to me', spoke volumes. In the published version of the paper whose draft was under discussion between them here, 'Wahrheit und Bewährung' (1936c), he accordingly declared his independence from both warring parties in the protocol sentence debate (without naming either Schlick or Neurath).[13]

As to the second difference between Schlick and Carnap, recall that already in his response to Schlick's earlier intervention, Carnap noted that it seemed to him that Wittgenstein allowed as analytic sentences only those for which one possesses an effective decision procedure. Since Carnap also referred to Schlick's 'The Foundation of Knowledge' in this context, Schlick responded that, since Wittgenstein's name was not mentioned there, nothing can be inferred from it about Wittgenstein's position.[14] Schlick's rejoinder is well-taken, but this leaves open the question of his own position. Consider the view of analyticity at issue.

In a section of *Logical Syntax* that had to be cut from the German version and was published separately in *Monatshefte für Mathematik und Physik* but reinserted in the English translation, Carnap wrote:

> When Wittgenstein says [*Tractatus*, p. 164]: 'It is possible [...] to give at the outset a description of all "true" logical propositions. Hence there can be no surprises in logic. Whether a proposition belongs to logic can be determined', he seems to overlook the *indefinite* character of the term 'analytic' – apparently because he has defined 'analytic' ('tautology') only for the elementary domain of the sentential calculus, where this term is actually a definite term. The same error occurs in Schlick ['Fundament', 1934, p. 96] when he says that directly a sentence is understood, it is also

> known whether or not the sentence is analytic. 'In the case of an analytic judgment, to understand its meaning and to see its *a priori* validity are one and the same process.' He tries to justify this opinion by quite rightly pointing out that the analytic character of a sentence depends solely upon the rules of application of the words concerned, and that a sentence is only understood when the rules of application are clear. But the crux of the matter is that it is possible to be clear about the rules of application without at the same time being able to envisage all their consequences and connections. The rules of application of the symbols which occur in Fermat's theorem can easily be made clear to any beginner, and accordingly he understands the theorem; but nevertheless no one knows to this day whether it is analytic or contradictory. (Carnap 1937, §34a, pp. 101–2; cf. Carnap 1935d, p. 167)

Against this, Schlick remarked in his letter to Carnap of 14 November 1935 that he retained the claim from 'Foundation' (here attacked by Carnap) still in its French translation 'Sur le fondement de la connaissance' (1935a), because he believed he had good reasons for it all along. 'I do believe, after all, that one can say of a mathematical proposition that one has understood it only if one has proved it. In order to elucidate this, I would have to range very widely and explain Wittgenstein's new ideas' (ASP, RC 102-70-11). Schlick did not elaborate, nor did Carnap ask him to do so in his response. Clearly, Carnap's ingenious method of accommodating the thesis that arithmetic is analytic to Gödel's incompleteness result did not find favour with Schlick (or Wittgenstein). Given the importance of this way of treating analyticity for Carnap's project in *Logical Syntax*, this is a major disagreement that tells conclusively against the idea that Carnap's understanding of logical tolerance was shared by Schlick (or Wittgenstein).

A third major disagreement arose over Carnap's claim that it is possible 'to construct a language with extra-logical rules of transformation', so-called P-rules (as opposed to purely logical or definitional L-rules). 'The first thing which suggests itself is to include amongst the primitive sentences the so-called laws of nature, i.e. universal sentences of physics ("physics" is here to be understood in its widest sense).' For Carnap, it was 'a matter of convention and hence, at most, a question of expedience' whether in the construction of a language one formulates only L- or also P-rules (Carnap 1937, §51, p. 180). So what Carnap suggested here was that it was possible to include synthetic sentences amongst the propositions that constitute a logico-linguistic framework within which scientific theories could be reconstructed. And he opposed this view to Wittgenstein's 'absolutist conception of language' in the *Tractatus* at 6.113 that 'it is a very important fact that the truth or falsity of non-logical propositions can *not* be recognized from the propositions alone'.

> It is certainly possible to recognize from its form alone that a sentence is analytic; but only if the syntactic rules of the language are given. If these rules are given, however, then the truth or falsity of certain synthetic sentences – namely, the determinate ones – can also be recognized from their form alone. It is a matter of convention whether we formulate only L-rules, or include P-rules as well; and the P-rules can be formulated in just as strictly formal a way as the L-rules. (Carnap 1937, §52, p. 186)

Carnap argued that, given the rules of language in question, we can also recognize the truth of certain synthetic sentences, for we can recognize them as framework propositions, namely, P-rules.

Schlick strongly opposed what he perceived this conception of Carnap's to be and devoted an entire paper at the Paris Congress, 'Are Natural Laws Conventions?' (1936), to oppose it, albeit so discreetly that Carnap did not realize it until he received Schlick's letter of 14 November 1935 prior to publication. The published version makes matters very clear since footnote 3 states:[15]

> It is true that a sentence (a sign sequence) which, under the presuppositions of customary grammar, expresses a natural law can be made into a principle of language simply by stipulating it as a syntactical rule. But precisely by this device one changes the grammar and, consequently, interprets the sentence in an entirely new sense, or, rather, one deprives the sentence of its original sense. It is then not a natural law anymore at all; it is not even a proposition, but merely a rule for the manipulation of signs. This whole reinterpretation appears therefore trivial and useless. Any interpretation which blurs such fundamental distinctions is extremely dangerous. (Schlick 1936/1979, p. 445)

In the letter mentioned, Schlick went still further and accused Carnap, in his preliminary draft of 'Wahrheit und Bewährung', of obliterating the difference between 'reality' and 'the description of reality' and suggesting that 'reality is created by language such that primitives and quantum physicists live in different realities'.[16]

Carnap responded by first pointing out that in his understanding too sentences were individuated by their rules of application, albeit their syntactical rules, and that 'besides the formal, syntactic statements about language it is also possible to make psychological, historical, sociological ones, etc.'.[17] In other words, Schlick seems to have misunderstood the reach of his formal analysis. (Schlick may well have been amongst the first who were fooled by Carnap's ostensive rhetoric that his logical syntax had nothing to do with meaning.) Moreover, Carnap added, 'the difference between reality and its description is a matter of course for me just as it is for you'. This did not settle the matter though, for there emerged another difference of opinion to which we will turn presently.

Carnap's P-rules have been criticized in a Schlickean spirit by Alberto Coffa. Coffa too was scandalized by Carnap's remarks about them in opposition to Wittgenstein. It seems he suspected Carnap of quietly harbouring a coherence conception of truth beneath his official syntactic rejection of truth as such and he claimed that 'the link between a synthetic sentence and its truth value is not up to us' (1991, p. 321). What Coffa overlooked, like Schlick, is that for Carnap, placing a law of nature amongst the P-rules for a given language did not mean declaring it true by convention, but rather placing it amongst the framework propositions of that language. Elsewhere Carnap noted explicitly that the truth of such synthetic sentences was determined independently.

> If P-rules are desired, they will generally be stated in the form of P-primitive sentences; we will call these primitive laws. [...] A sentence of physics, whether it is a P-primitive sentence, some other valid sentence, or an indeterminate assumption (that is, a premiss whose consequences are in course of investigation), will be *tested* by deducing consequences on the basis of the transformation rules of the language, until finally sentences of the form of protocol sentences are reached. These will then be compared with the protocol sentences which have actually been stated and either [be] confirmed or refuted by them. If a sentence which is an L-consequence of certain P-primitive sentences contradicts a sentence which has been stated as a protocol sentence, then some change must be made in the system. For instance, the P-rules can be altered in such a way that those particular primitive sentences are no longer valid; or the protocol-sentence can be taken as being non-valid; or again the L-rules which have been used in the deduction can also be changed. There are no established rules for the change which must be made. (Carnap 1937, §82, p. 317, italics in original)

What Carnap gave expression to here, of course, was a variation on Duhem's famous claim that faced with what appears to be contrary evidence for an underdetermined claim, logic alone will not tell us what to do (but only '*bon sens*'). Clearly, rendering putative laws of nature as P-rules did not make them true by convention, full stop. 'No rule of the physical language is definitive; all rules are laid down with the reservation that they may be altered as soon as it seems expedient to do so' (ibid., p. 318). Carnap, I submit, is not guilty of the conventionalist crime alleged by Coffa and Schlick.[18]

So is Schlick's criticism of Carnap's conventionalism just based on a misunderstanding? It would appear that it was not. Here we come to the fourth difference between them. Behind Schlick's criticism stood the firm conviction that 'we should like to mean by "natural law" something that is invariant relative to any arbitrary mode of formulation. And this is possible' (1936/1979, p. 443). For Schlick, facts and conventions were always sharply separable. As he stressed in his letter of 14 November to Carnap,

> Talk of *the* reality points to the invariance which consists in that, if one possesses a description of the world in any one language, one is able, by means of purely grammatical transformations, to produce the correct description of the world in any other language. It is arbitrary whether I describe the star system in Euclidean or non-Euclidean terms; but after I have determined, e.g., to consider light rays as 'straight lines' I am bound, I have made the Euclidean description impossible; via transformations, however, I can always change over to such a description. (Schlick to Carnap, 14 November 1935; ASP, RC 102-70-11)

Precisely this, of course, is what Carnap denied: 'I do not believe in translatability without remainder and I think therefore that also the content of the description of the world is influenced to a certain degree by the choice of the form of language.'[19] In the published version of 'Wahrheit und Bewährung', Carnap gave the following example:

> The answer to a question concerning reality however depends not only upon that 'reality', or upon the facts but also upon the structure (and the set of concepts) of the language used for the description. In translating one language into another the factual content of an empirical statement cannot always be preserved unchanged. Such changes are inevitable if the structures of the two languages differ in essential points. For example: while many statements of modern physics are completely translatable into statements of classical physics, this is not so or only incompletely so with other statements. The latter situation arises when the statement in question contains concepts (like, e.g., 'wave-function' or 'quantization') which simply do not occur in classical physics; the essential point being that these concepts cannot be subsequently included since they presuppose a different form of language. This becomes still more obvious if we contemplate the possibility of a language with a discontinuous spatio-temporal order which might be adopted in a future physics. Then, obviously, some statements of classical physics could not be translated into the new language, and others only incompletely. (This means not only that previously accepted statements would have to be rejected; but also that to certain statements – regardless of whether were held true or false – there is no corresponding statement at all in the new language.) (Carnap 1936c/1949, p. 126)

Clearly, what Carnap gave expression to here was the phenomenon of conceptual incommensurability between different scientific theories. Schlick was led into denying this incommensurability by his opposition to Carnap's perceived violation of the grammar of talk of natural laws. Schlick either missed or rejected the central point of *Logical Syntax*, what Michael Friedman has called its reconstitution of the notion of a relative a priori.[20] His further correspondence with Carnap did not return to the matter, so it must remain open what he made or would have made of Carnap's extended published argument on further reflection. (In his letter of 4 December 1935, Carnap thanked Schlick for having prompted him to set out his thinking on the matter much more carefully. Whether Schlick saw the published version of 'Wahrheit und Bewährung' is not known; he was murdered half a year later.)[21]

There are reasons to think that Schlick's reaction was not out of character. Already back in 1920 in his exchange with Reichenbach on the latter's first book on the theory of relativity, Schlick showed little patience with the idea of a relative a priori and instead convinced his interlocutor to drop such talk in favour of a sharp distinction between conventionally determined coordinating definitions and experimentally determined factual assertions. Whatever else may be said of his rejection of Carnap's empiricist-conventionalist revitalization of Reichenbach's updating of the Kantian a priori, Schlick had form.[22]

Lastly, there is a candidate for a fifth difference between Schlick and Carnap. Direct evidence for this is hard to come by, but were this to hold, it would also connect with an earlier stance of his. In 'Facts and Propositions', while claiming to respect and agree with the distinction between the formal and the material mode of speech, Schlick again made moves that undermine this impression. Thus he denied that the formal mode is 'more correct' and claimed that the material mode of speech is 'as such not faulty'. This suggests that Schlick was not prepared to accept the object-/metalanguage distinction that was the primary object of drawing the formal mode/materials mode distinction.[23] (Was it perhaps Wittgenstein's animadversions against 'Metalogik' in the *Big Typescript* that held Schlick prisoner in 1934, just as in 1930 he was held prisoner by Waismann's 'Theses'?)

Even more than Schlick's objections to incommensurability, this objection would go to the very heart of *Logical Syntax*. In so far as the distinction between the material and formal modes expressed at bottom nothing more than the distinction between object- and metalanguage (the exclusion of semantics being a merely temporal accretion), it is precisely this distinction that codifies Carnap's transgression of Wittgenstein's known strictures against metalinguistic discourse in the *Tractatus*, strictures that were perceived to have been reasserted around 1930 when Waismann's 'Theses', in their first introductory sentence, declared itself to consist only of 'elucidations'. But even if Schlick's criticism did not go quite as far, it seems clear enough that he was out of sympathy with Carnap's project of developing philosophy as logic of science. His conception of the new philosophy after the 'turning point' was more akin to developing, in what he took to be Wittgenstein's fashion, the 'grammar' of ordinary language and experience.

## 5 Neurath's favourable reception of *Logical Syntax*

Turning to Schlick's opponent in the protocol sentence debate, Neurath, we find praised just those doctrines that prompted Schlick to dissent. Now, especially as regards the undercover semantics of *Logical Syntax*, Neurath's positive reaction is not without problems of its own, but since these lead way beyond into Carnap's semantic phase I cannot discuss them here.[24]

In early March 1934, when Neurath broke his return journey from Moscow to Vienna in Prague on account of the threat of imprisonment in Vienna (he never returned there), Carnap gave a copy of the typescript to Neurath to read. The reaction was wholesale but enthusiastic:

> Dear Carnap! As is my way, I first leafed through your book in order to grasp its construction and train of thought and in order to see whether the lovers find each other. As always, so this time too I see with great admiration how everything comes together in a consistent whole.... about details I want to talk to you and Frank at greater length. (Neurath to Carnap, 8 March 1934, ASP, RC 029-10-91)

One month later, Neurath was on a boat sailing from Danzig (Gdansk) to his first exile in The Hague.

> With greatest delight I have now been reading, on deck and in bed, your *Logical Syntax* with greater care. I hope that the third and fourth reading will seriously acquaint me with the details. I believe that this [book] is a great step forward [...]
>
> I am wondering which views of my own I must revise now. Important the conventionalistic conception of the *choice* between languages, in place of talk of 'the' language. Problem, which difficulties remain? Meaningless expressions can be avoided by a good syntax *ad limine* [...] but there remain the wolves which one has locked in the pen unknowingly. (Neurath to Carnap, 5 April 1934. ASP, RC 029-10-77)

This positive evaluation remained despite the reservation noted. Later that year, Neurath wrote to Carnap requesting to buy another copy at author's price from him, because he had lost, as he put it, his 'logical bible'.[25]

Earlier on, before the book went to press, it seems that Carnap asked Neurath about suggestions for the Foreword. (At roughly that stage, Neurath also asked for, and received, citation of his work on physicalism.)[26] In any case, Neurath did offer such advice.

> Dear Carnap. In a rush. To your Foreword ... And the ending of the Foreword could now easily be taken from S 3. If you do not wish to end with a fanfare, then at least with the melodic song of yearning of a Prague nightingale. S 3 'The present investigation concerns ...' is, as a fanfare, too bureaucratic and, for a nightingale, too stiff. Rather like this:
>
> 'Task of fundamental importance!'
>
> 'Timid hesitation ... everyone who seeks out new paths requires justification.'
>
> 'Attempts to leave the coastal waters of classical logic.'
>
> '(Ahead of us the wide blue distance), *the boundless ocean of possibilities*.'
>
> That would be a nice ending. It would correspond to the ending in *Logische Aufbau* and would not make difficulties for your search for a chair. (Neurath to Carnap, 10 June 1934. ASP, RC 029-10-65)

Carnap, as readers will have realized, was happy to accept some of these suggestions. And while this cannot be proved without access to Carnap's initial draft for the Foreword (which seems to exist no longer), the very tenor of Neurath's writing on this occasion suggests that Carnap's boat simile in *Logical Syntax* was suggested by none other than Neurath himself!

Carnap responded:

> I accepted several of your suggestions for changes in the Foreword. Unfortunately, not all. The references to the literature and the names of the authors I was unable to shift into the Appendix. Instead I separated the 'pretty' part of the Foreword from this sober one by a line. The Foreword now consists of two parts. (Carnap to Neurath, 18 May 34. ASP, RC 029-10-63)

That line across the page did indeed appear in the original of *Logical Syntax*, but was dropped in translation; moreover, the book carries no Appendix. Not surprisingly, one of the pieces of advice Neurath gave was not to foreground Wittgenstein's early influence too much; whether Carnap responded to this cannot be told, but as we noted earlier, on that point Schlick's advice went in the opposite direction. What may be noted though is that the passage that concluded the 'pretty' part that in the English translation became four paragraphs was just one long paragraph in the German original. That paragraph covered the breadth of Neurath's advice and ran from the claim that, while Carnap's sketch of a general logical syntax represented but an early attempt, its task remained 'one of fundamental importance', through its introduction of the principle of tolerance as the solvent for the foundational dispute in mathematics, all the way to the adoption of Neurath's suggestions for a maritime simile. In the English edition, the 'pretty' part of the Foreword now concluded with the following paragraph:

> The first attempts to cast the ship of logic off from the *terra firma* of the classical forms were certainly bold ones, considered from the historical point of view. But they were hampered by the striving after 'correctness'. Now, however, that impediment has been overcome, and before us lies the boundless ocean of unlimited possibilities. (1937, p. xv)[27]

Critical tones only enter the Carnap–Neurath correspondence (with the exception of an exchange about the merit of Carnap's positive evaluation of Popper's proposal for protocol statements) after the Paris Congress. Not surprisingly, perhaps, they concern Carnap's reintroduction of the concept of truth into the logic of science (following Tarski's work). Since in *Logical Syntax* the concept of truth was officially banished, one can view Neurath's concern about Carnap's semantic turn as an indication of how deeply he agreed with what he took *Logical Syntax* to be saying when it rejected the use of the concept of truth as a non-syntactic concept. The issue of their dispute about semantics is too far-reaching to be discussed here, but it may still be of interest to note how, in its early stages, the matter was viewed by Neurath.

While applauding Carnap's response to Schlick's demand for univocality and correspondence truth without remainders, Neurath was worried by Carnap's new stance on truth and soon he characterized his dissent from the semantic project as follows:

> I believe that Tarski (and you too) will start a lot of confusion with the remarks which are correct by themselves. He shows that the concept of truth that is used in everyday speech cannot be used in everyday speech, but only in formal languages. So just where one does not need this term but could replace it by another arbitrary one. Continuity is not preserved. [Tarski] is looking only for a formal model that satisfies certain conditions. I am looking for a concrete mass of sentences, for the comprehension of which one employs formal models. (Neurath to Carnap, 23 December 1935. ASP, RC 102-50-01)

But with this Neurathian emphasis on what one may call the pragmatics of science – as opposed to the logic of science – I shall leave the matter. The potential of combining Neurathian pragmatics and Carnapian logic of science remained historically unfulfilled but systematically intriguing.[28] Instead, Neurath's celebration of *Logical Syntax* marked the high point of their collaboration.

## 6 Hahn's anticipation of logical pluralism

The overall agreement with *Logical Syntax* that Carnap found on the left wing of the Circle was not confined to Neurath. Even though Hahn died unexpectedly while *Logical Syntax* was still in press and no clear endorsement of it by him exists, it is possible to show that he was wholly in sympathy with it. Hahn made important but widely unrecognized moves that must count as distinctive moves in the direction of logical tolerance and pluralism in the Carnapian understanding of the term.

It is not uncommon when Hahn's philosophy of mathematics is discussed in the secondary literature for him to be characterized as a more or less typical logical empiricist: getting his ideas from Wittgenstein but mixing them up in the process. Something important is lost in this perspective: precisely his early moves towards logical tolerance. Establishing this is not an easy matter, however, and for details I must refer readers elsewhere (Uebel 2005). That said, two points are particularly noteworthy and may be mentioned here.

First, Hahn too started from but extended Wittgenstein's concept of tautology. By rejecting Russell's conception of logic, Wittgenstein effected an advance which Hahn summarized as follows:

> If logic were to be conceived – as it has actually been conceived – as a theory of the most general properties of objects, as a theory of objects as such, then empiricism would in fact be confronted with an insuperable difficulty. But in reality logic does not say anything at all about objects; logic is not something to be found in the world; rather, logic first comes into being when – using a symbolism – people *talk about the world*, and in particular, when they use a symbolism whose signs do *not* (as might at first be supposed) stand in an isomorphic one-one relation to what is signified (the introduction of a symbolism by means of an isomorphic one-one projection would be of little interest). (1929/1980, p. 40)

Consequently, Hahn noted, the law of contradiction 'does not say anything about the world but rather deals with the way in which the symbolism used is supposed to *designate*' (ibid., p. 41). Already in 1929 Hahn also called logic a 'set of directions for making certain transformations *within the symbolism we employ*' (ibid., p. 24, italics added). The bare hint given here that the symbolism 'we employ' may take variable logical forms was not yet further explored, however.

In lectures of the year 1932 Hahn repeated his claim that logic 'deals only with the way we talk about objects; logic first comes into being through language' (1933a/1987, p. 29). But this time he also signalled explicitly (but still discreetly in a footnote) that his ideas went beyond Wittgenstein's, though again he lauded Wittgenstein's 'decisive contribution to the line of thought developed' in his lectures (ibid., n. 11).[29] In Hahn's view, tautologous statements 'merely express a dependence in the assignation of designations to objects' (ibid., p. 32). By contrast, he noted that Wittgenstein used the term in a 'narrower sense', namely, as 'true merely by virtue of its form' (ibid., n. 11). Now what made Hahn's use wider?

Hahn aimed to replace Wittgenstein's transcendentalist conception of logic with a conventionalist one. For Hahn, Wittgenstein's conception of truth in virtue of form did not simply draw a contrast with truth in virtue of fact, but it also represented a very specific understanding of what logical form amounted to. In the *Tractatus*, the formal was not characterized simply by the Hilbertian abstraction from all content, but also by its 'transcendental' function: 'Logic is not a body of doctrine, but a mirror-image of the world. Logic is transcendental' (Wittgenstein 1922, 6.13). The conception of logical form as what a representation 'must have in common with reality in order to be able to depict it – correctly or incorrectly – in any way at all' (ibid., 2.18) was for Hahn an essential aspect of Wittgenstein's advance in thinking of logic as tautologous. But that advance was only partial. It still bore traces of the conception it aimed to replace: even if logic no longer established the most general truths about objects in general, it did still reflect the 'formal – logical – properties of the world' (ibid., 6.12). For Hahn, this was a 'narrow' conception of tautology as opposed to a 'wide' one which would free logic from the 'transcendental' office of 'mirroring' any properties of the world. Hahn's intention was the articulation of such a wider conception. The central idea implied in his holding that logic concerns the way 'we *want* to talk' about the world (1933a/1987, p. 31, italics added) – namely, that fixing the logico-linguistic framework was or could be part of the human repertoire – was not something acknowledged in Wittgenstein's scheme of things in the *Tractatus*.

Reading Hahn's remarks on tautologies against the background of the opposition by some members of the Vienna Circle to correspondence conceptions of truth and knowledge, their intended point is thrown into clear relief. First and foremost, Hahn was concerned to free Wittgenstein's conception of tautology from what he (and, notoriously, Neurath) perceived to be the metaphysical trappings of the *Tractatus*, its correspondence-theoretical conception of truth and language. Did Hahn not write that 'it is a big mistake to infer the structure of the world from the structure of language' (1930b/1980, p. 8), even that '[l]ogic arises when and only when what is to be depicted and its pictures, the symbols, exhibit different structures' (1930a/1980, p. 24)?[30] Yet, second, in the advance beyond Wittgenstein's perceived correspondentism, it was also the notion of *the* logic that increasingly came under pressure. So Wittgenstein's conception of tautologies was 'narrower' in that truth in virtue of logical form remained a fact beyond human meddling. By contrast, for the price of a conventionalist-pragmatic story still to be told, in Hahn's 'wider' conception such truths depend on the representational system adopted. Hahn's tautologies 'say nothing about objects but merely lay down rules about how we want to talk about objects' (1933a/1987, p. 31). Hahn here articulated the doctrine of logical truth as truth by convention. What he barely hinted at in 1929 was clearly in evidence by 1932.

The second point to note about Hahn's philosophy of mathematics is that in the dispute about the foundations of arithmetic, Hahn both declared himself for a logicism that still awaited its final formulation *and* began to revise the idea of what counted as a resolution of the dispute. At the Prague conference in 1929 Hahn merely stated that Russell's articulation of logicism in *Principia* stood in need of reform. Just what that reform amounted to he specified at the Second Conference for the Epistemology of the Exact Sciences in Königsberg one year later. Noting again that the task was not yet fully accomplished, Hahn stated that 'the formal side' of Russell's system was 'largely in order as it is and highly suitable for the foundations of mathematics' (1931/1980, p. 35), but he objected to Russell's 'absolutist-realist' or 'realist-metaphysical position' (ibid., pp. 34–5). Of course, the problem of the axioms of choice and infinity needed fixing (besides making the switch back to the simple theory of types so as to avoid the need for the axiom of reducibility) before Russell's system could be considered fully in order. Yet what Hahn was concerned to stress was that 'a different philosophical interpretation' of the system was needed (ibid., p. 35).

Hahn's solution to both the technical and the philosophical problem is contained in the following passage outlining his preferred interpretation of the logicist reconstruction of arithmetic.

> I assume, like Russell, that for describing the world (or better: a section of the world) we have at our disposal a system of predicative functions, of predicative functions of predicative functions, etc. – though, unlike Russell, I do not believe that the predicative functions are something absolutely given, something we can point out in the world. Now the description of the world will turn out differently according to the richness of this system of predicative functions; we therefore make certain *assumptions* about its richness [...] Now the whole of mathematics arises out of the tautological transformation of the requirements we make about the richness of our system of predicative functions. Whether a certain proposition is or is not valid [...] depends on the requirements we have made about the richness of the underlying system of predicative functions, or if you want to call them that, on the *axioms*; the question about the *absolute* validity of such propositions is completely senseless. (1931/1980, pp. 35–6)

The key word is 'assumption'. Here Hahn's preferred response to the problematic axioms of choice and of infinity shines through: Russell's own, as best remembered from his *Introduction to Mathematical Philosophy*.[31] Being only 'necessary for certain results but not for the bare existence of deductive reasoning', these axioms 'could perfectly well be stated as an hypothesis whenever [they] are used, instead of being assumed to be actually true' (1919, p. 191). Thus they are not asserted as independent axioms – for they 'cannot be asserted by logic to be true' (ibid., p. 203) – but are appended to mathematical statements dependent on them as their conditional antecedent. With the mathematical theorems in question conditionalized on the problematic axioms in this way, the resultant mathematical statements became logical truths.[32] However, this technical solution to the problems of logicism brought philosophical changes in its train, as is easily seen. The derivation of arithmetic can no longer be claimed to be effected from logic alone but only from logic plus certain assumptions: the derivation was no longer absolutely valid but only relative to those assumptions. Moreover, reconstructed in this way, realism lost its grip, for any such reconstruction of mathematics could not shake off its dependence on these assumptions.

Now importantly, in Hahn's derivation of mathematics from logic the absolutism of supposing one system of predicative functions etc. to represent *the* logic was overcome. Precisely because of it, a solution to the problem of the foundation of transfinite arithmetic seemed to be within reach: 'the requirement that the axioms of infinity or the axiom of choice be valid is in this sense a requirement about the richness of the system of predicative functions by means of which I want to describe the world' (1931/1980, p. 36). The difference from Wittgenstein could not be clearer: 'If we know the logical syntax of any sign-language, then we have already been given all the propositions of logic' (1922, 6.124). By contrast, for Hahn, if we know the logical syntax of a given language, we have been given only one possible logic.

With his remarks at the Königsberg discussion in 1930, then, Hahn introduced a sharp break with Frege, Russell, and Wittgenstein. Consistent with the conventionalism he harboured since 1929, quite different 'assumptions about the richness' of the languages in question could be made. In consequence, not only did it no longer make sense to speak of the 'absolute validity' of the axioms adopted, it also no longer made sense to speak of one universal logic determining 'the way in which we speak about objects'. Already Hahn's approach was that of a language constructor or engineer who – unlike Wittgenstein who transcendentalized logic – assumed a version of the idea of logical pluralism (without, to be sure, either calling it by this name or even designating it as a principle).[33] Hahn's remarks also gave a new twist to the foundational dispute between logicists, formalists, and intuitionists. Once a quest for the ultimate foundation of mathematical truth (be it logical, metamathematical, or *ur*-intuitive), in Hahn's hands it became a pragmatic issue. It concerned the question of what the requirements of the richness of language would be that suffice for the reconstruction of mathematics of varying strengths; differently put, it concerned the choice between different sets of criteria of adequacy for the analysis of the language of arithmetic.

By way of a brief chronology, let us note that Hahn was evidently thinking along the lines of logical pluralism at least since September 1930. Carnap's contribution to the Königsberg discussion did not yet embrace that idea, his concern being focused on the equivalence of logicist and formalist constructions of mathematics via the investigation of the applicability of mathematics.[34] In fact, we must not be misled even to think of the start of the metalogic project in January 1931 as the beginning of Carnap's logical tolerance. He did so only by placing the option for Language II alongside that for Language I – after, that is, he had been confronted by Hahn's and Gödel's criticisms in July 1931 that the still universal Language I of his later *Syntax* (the only one he explored in his three metalogic lectures to the Circle during the previous month) was too restrictive with regard to the mathematics it allowed.[35] So while Carnap and Hahn developed their non-standard logicist ideas vaguely in parallel around 1929–30, Hahn did possess a distinctive conventionalist edge which found its clearest expression at Königsberg. Moreover, it was Hahn who raised the conventionalist objections to the metalogic lectures that – together with Gödel's objections one year later to Carnap's first definition of analyticity[36] – set their author Carnap free from the *terra firma* of mathematical foundationalism to pursue the path of logical pluralism of *Logical Syntax*.[37]

## 7 Conclusion

Celebrated and criticized as the conclusion of a monumental effort to overcome the foundational impasse in the philosophy of mathematics whose first steps had been witnessed in his metalogic lectures to the Circle in late spring of 1932, *Logical Syntax* documented and confirmed the divisions that had come to obtain in the Circle. While Schlick elaborated and hinted at objections that suggest a Wittgensteinian provenance, other colleagues were much more supportive: not only Neurath with his wholesale enthusiasm, but also Hahn with his anticipatory remarks at Königsberg and his comments on Carnap's metalogic lectures. Had both Hahn and Schlick survived longer, this contrast between the conservative and the left wing of the Circle surely would have found far more public expression than it did as a matter of historical record, taking *Logical Syntax* as its focal point.

## Notes

1. Unless noted otherwise, in this chapter emphases in quotations are from the original.
2. Menger to Carnap, 3 February and 15 March 1934; Carnap to Menger, 21 March 1934 (ASP, RC 029-01-13, 029-01-09, 029-01-08). The quotation is from the second of these letters. (Compare also Köhler 1991.) Incidentally, the bibliography of Carnap (1934d/1937) lists Menger (1928) and (1933), which were not mentioned in the text, alongside Menger (1930) which was. Unless otherwise noted, in this paper translations from archive sources are by the present author.
3. In the letter of 15 March 1934, Menger criticized an earlier form of the acknowledgement in proof, having more or less subtly demanded one in his previous letter of 3 February (see note above). Menger especially opposed Carnap's mention of Gödel in this connection (which Carnap then dropped), correctly so as Gödel's pluralism, his acquiescence in an infinity of mathematical systems, is quite tangential to the claim that there is no 'true logic' or privileged understanding of constructivity; see Köhler (1991 and 1993).
4. See Carnap (1930b), (1931c), (1934d/1937, §84), and (1963a, p. 48). For discussion see Richardson (1994) and Uebel (2005).
5. In his memoirs, Menger noted that he gave a report on his 'Bemerkungen' to the Circle but did not specify a date and portrayed his opposition to talk of 'the' language and 'the' logic as ongoing (1994, p. 200). Menger's remarks on another occasion (1979a, p. 12) suggest the year 1927/28; still elsewhere he noted that he gave a separate presentation during the academic year 1928/29 in the course of which he criticized talk of 'the' language and/or logic (1982, p. 89).
6. On the dating of Carnap's embrace of logical pluralism, see also the last paragraph of section 6 below.
7. See Menger (1994, pp. 148 and 156) and Carnap (1963a, p. 30).
8. Menger noted that Schlick's reaction to the lecture of Tarki's he attended was 'a bit cool' (1994, p. 151) and that in the discussion that followed he tended to side with Waismann's Wittgensteinian objections to metamathematical and metalinguistic discourse in general (1982, p. 94); he blamed Wittgenstein's influence for having created in Schlick and Waismann 'a bias against formal logic and mathematics' (ibid., p. 95). Carnap noted Schlick's 'scepticism' with regard to the philosophical usefulness of the metamathematical method (1963a, p. 30).
9. On the earlier priority dispute between Wittgenstein and Carnap, see Uebel (1995) and Stern (2007).
10. The published German original reads: 'Zu meinen Bemerkungen, besonders in §§17 und 67, gegen Wittgensteins frühere dogmatische Einstellung teilt mir jetzt Herr Schlick mit, dass Wittgenstein schon seit mehreren Jahren in unveröffentlichten Arbeiten die Regeln der Sprache als völlig frei wählbar darstellt' (Carnap 1934d, pp. vi–vii). Note that Carnap preferred 'some time' in the translation to the 'mehrere Jahre' in the original.
11. A copy of the typescript is preserved amongst Carnap's papers (ASP, RC 102-76-02) with the handwritten title (most likely by Carnap) '*Waismann* (Thesen aus dem Zirkel, 1930)'. It has since been published as 'Appendix B' in McGuinness (1967/1979, pp. 233–62).
12. McGuinness (1967/1979, §§7–8). For a similar view on laws of nature as hypotheses which is credited to Wittgenstein, see Schlick (1931/1979, p. 188).
13. (ASP, RC 102-70-10).
14. Schlick to Carnap, 5 June 1934 (ASP, RC 029-28-10).

15. It would seem that footnote 3 in Schlick (1936), which identifies the Carnapian opposition, was not contained in the manuscript which Feigl read for Schlick at the Congress.
16. Schlick to Carnap, 14 November 1935 (ASP, RC 102-70-11).
17. Carnap to Schlick, 4 December 1935 (ASP, RC 102-70-10).
18. Still more recently the Schlick–Coffa criticism has been recast by Tom Oberdan as expressing, in misleading ways, the supposedly correct insight that Carnap's concept of P-rules as contributing to the definition of a language should be dropped for they do not contribute to the constitution of meaning – which alone is claimed to be the proper domain for the philosophical principle of logical tolerance (2004, p. 136). On Coffa's claim see also Goldfarb (1997).
19. Carnap to Schlick 4 December 1935 (ASP, RC 102-70-10).
20. See Friedman (1994) and (2000).
21. Compare Oberdan (1993, pp. 139–41).
22. And we may note that it has recently been argued forcefully that Schlick's preferred form of moderate conventionalism did not do justice to the facts of the general theory of relativity as it claimed to do: according to that theory it is precisely not an arbitrary but an empirically determined matter whether the star system – physical geometry – is described in Euclidean or non-Euclidean terms. See the discussion in Ryckman (1992, 2005).
23. The distinction between the material and formal mode of speech applies to philosophical discourse after the linguistic turn. In essence it says that philosophy qua metatheory speaks about the statements made and linguistic expressions used in first-order inquiries and not about the objects of first-order inquiries themselves. The so-called formal mode of speech seeks to make this metalinguistic nature of philosophical talk explicit, whereas the common so-called material mode of speech does not draw the required distinction and leaving it at best implicit is liable to mislead into philosophical pseudo-questions. It was an inessential accretion during Carnap's syntactic phase that the formal mode of speech barred talk of meaning.
24. There obtains one puzzle concerning Neurath's opposition to semantics. Neurath loved *Logical Syntax* but hated Carnap's acceptance of Tarskian truth. Didn't he notice the undercover semantics that *Logical Syntax* contained? Why did Neurath accept the notion of logical truth as a (supposedly) syntactic concept, but not as a semantic one? The answer would appear to lie in the fact that Neurath suspected Carnap's semantics of more metaphysics than his disquotational approach to truth actually contained. See also Mancosu (2008b).
25. Neurath to Carnap, 25 November 1934 (ASP, RC 029-10-07).
26. See Neurath to Carnap, 8 March 1934 (ASP, RC 029-10-91): 'When I continue to write about the matter, I want to appear as somebody who continues his own work, not as someone who deviates from your line.' Sometime later, by numeration before the end of April, Carnap sent Neurath what appears to be copy of a page of typescript correction, an insert for §82 ('Fahne 154') which contains (except for one word) the literal version of the published passage wherein Carnap discusses as pathbreaking Neurath's work on physicalism (ASP, RC 029-10-73). One may see in this a continuation of the default compromise of their priority dispute about physicalism; see Uebel (1992, 2007 ch. 8).
27. The 1934 original German reads as its last three sentences (not separated as a paragraph of their own): 'Jene ersten Versuche, das Schiff der Logik vom festen Ufer der klassischen Form zu lösen, waren, historisch betrachtet, gewiss kühn. Aber sie waren gehemmt durch das Streben nach "'Richtigkeit". Nun aber ist diese

Hemmung überwunden; vor uns liegt der offene Ozean der freien Möglichkeiten' (1934d, p. vi).

28. See Uebel (2001) and (2007a, ch. 12).
29. Beyond his pointer to his difference from Wittgenstein, what Hahn actually says in his (1933) lectures, given in spring and autumn of 1932, concerning logic and mathematics does not go beyond his earlier pronouncements in (1929), (1930a), and (1931).
30. Whether Wittgenstein himself actually intended such a correspondentism is another matter, but it can hardly be claimed to be an outlandish interpretation of the picture theory of the *Tractatus*.
31. In *Introduction*, Russell made this suggestion for all three problematic axioms, but particularly for that of reducibility. In *Principia*, the latter axiom was still asserted boldly (Russell and Whitehead 1910–13, I, pp. 166–7) while the axioms of choice and infinity were already stated only hypothetically (ibid., II, pp. 101 and 203). Much later Russell recalled about the axiom of choice: 'We found no arguments either for or against this axiom, and we therefore included it explicitly in the hypothesis of any proposition which used it' (1959, p. 93). Except for new ways of dealing with the axiom of reducibility, these matters were not discussed in the *Introduction* and Appendices to the 2nd edition of *Principia*.
32. Carnap also accepted Russell's proposal in (1930b, p. 308) and (1931a, p. 96).
33. It points in the same direction that in a review of Kaufmann (1930), Hahn noted Kaufmann's 'philosophical' style and his desire to establish his finitist standpoint 'as the only possible one' and expressed the view that 'in many respects this book will not convince mathematicians'. See *Literaturberichte* of *Monatshefte für Mathematik und Physik* 38 (1931), pp. 6–7.
34. Menger once reported (1979a, p. 15, n. 3) that he had planned to present his logical pluralism at the Königsberg conference but was dissuaded from sending a deputy when he could not attend by Felix Kaufmann. It may be added that, like Carnap and unlike Menger, Hahn's philosophy of mathematics was also concerned with the problem of accounting for the applicability of mathematics.
35. See Hahn's remarks of 2 July 1931 in Stadler (1997/2001, pp. 297–8) and Gödel's comment of 12 July 1931 in Köhler (1991, p. 146). Hahn objected that 'the problem of the foundation of mathematics [...] only begins with the real numbers' to which Carnap conceded that he had so far accounted only for the natural numbers. Gödel made the point that since Language I can prove its own non-contradictoriness, given his own incompleteness proof, it must be a very impoverished language unable to express all of classical mathematics.
36. See Gödel (2003, pp. 346–57) and Goldfarb (2003).
37. Note also that Hahn once stressed – against Schlick's 'The form of the facts is mirrored in the language' – that 'There is no connection here unless it is artificially constructed. For the rules of syntax a logical justification cannot be given, because *it is only there that logic begins*' (19 February; in Stadler (1997/2001, p. 253, italics added)). Hahn's rejection of the idea of a pre-given logical form left the very distinction of form and content open for conventional determination.

# 2

# From Wittgenstein's Prison to the Boundless Ocean: Carnap's Dream of Logical Syntax

*Steve Awodey and A. W. Carus*

*The Logical Syntax* is a revolutionary book. How did the author of the *Aufbau*, whose viewpoint is so very different, come to write such a book? It was a drama in two acts, comprising not one but *two* major breakthroughs within less than two years. The first of these, in January 1931, was the one Carnap describes vividly in his autobiography:

> After thinking about these problems for several years, the whole theory of language structure and its possible applications in philosophy came to me like a vision during a sleepless night in January 1931, when I was ill. On the following day, still in bed with a fever, I wrote down my ideas on forty-four pages under the title 'Attempt at a Metalogic'. These shorthand notes were the first version of my book *Logical Syntax of Language*. (Carnap 1963a, p. 53)

The second, in October 1932, was Carnap's arrival at the principle of tolerance, the aspect of the book that has generated much of the recent interest, as other contributions to this volume make clear. This second idea was absent from the book's first draft, which focused on the initial 'vision' of January 1931; indeed, the principle of tolerance would appear to be in direct conflict with the main theme of this first draft.

In the published book of 1934 and in Carnap's mind then, these two ideas were fused into a single doctrine. This has led to much confusion about *Logical Syntax*, and the role of its ideas in Carnap's subsequent development. In this chapter we seek, therefore, to distinguish these two components of the book, to show how the first motivated the second, and explain which elements of the first step remained after the 'syntax' doctrine as a whole was discarded, within a year of the book's publication. The principle of tolerance, of course, became central to Carnap's entire subsequent development (Carus 2007).

Sections 1–5 below address the question we began with above: how did the author of the *Aufbau* get to step one?[1] Section 6 describes this step, and sections 7–8 discuss how Carnap got from step one to step two. Section 9, finally, shows how this new account of the genesis of the *Syntax* can be put to use in understanding not just the *Syntax* itself, but also its later traces, e.g. in the work of Quine.

One obstacle in talking about Carnap's pre-*Syntax* view is the large role played in it by certain ideas from Wittgenstein's *Tractatus*. Because the *Tractatus* has attracted so much more interpretive interest than the Vienna Circle, their understanding of the book has – we think somewhat unfairly – been classified as either a pale shadow or as an outright misunderstanding of Wittgenstein. We think this traditional view[2] misleading because it takes for granted that the Vienna Circle's first priority was a correct understanding of Wittgenstein's own intentions. In fact, of course, the Vienna Circle used whatever came to hand to address *their* problems, which were not Wittgenstein's. But his ideas, in their version, played a major role for them in the late 1920s.

## 1 The significance of the *Tractatus* for the Vienna Circle

Why was the *Tractatus* so important for the Vienna Circle? Because Wittgenstein had, in their view, solved the old Platonic problem of the cognitive status of mathematics, which was obviously a basic obstacle to any form of empiricism. 'It really does seem on first sight,' Hans Hahn said, 'as if the very existence of mathematics must mean the failure of pure empiricism – as if we had in mathematics a knowledge about the world that doesn't come from experience, as if we had a priori knowledge. And this evident difficulty for empiricism is so obvious that anyone who wants to hold a consistent empiricism has to face this difficulty...' (Hahn 1929/1988, pp. 55–6). Wittgenstein had solved this problem. Of course he was not an empiricist, so it had been no part of his agenda to solve this basic problem for empiricism. But the Circle took their solutions where they could find them.

The key idea was the picture theory of meaning. Language represents the world by isomorphically corresponding to the arrangement of its elements, i.e. giving a 'logical picture'. Atomic sentences picture atomic facts, and all other (meaningful) sentences are truth-functional concatenations of atomic sentences. Though Wittgenstein adopted Frege's and Russell's all-encompassing conception of logic as universally applicable and inescapable, he rejected their view that the logical laws were laws *of something* in the world (something like the most general laws of nature, or the laws of thought). For Wittgenstein, logical laws weren't laws *of* something; they were, rather, an artefact or by-product of isomorphic representation. Certain concatenations of propositions come out true (or false) regardless of what facts hold; these are 'tautologous' (or contradictory) and empty. They say nothing whatever about the world. The Vienna Circle thought this idea of critical importance:

> If one wants to regard logic – as this has in fact been done – as the study of the most general qualities of objects, as the study of objects in general [*überhaupt*], then empiricism would in fact be confronted here with an impassable hurdle. In reality, though, logic says nothing whatever about objects. Logic is not something that is to be found in the world. Logic only arises, rather, when – by means of a symbolism – we *speak about the world*... The sentences of logic say nothing about the world. (ibid., pp. 56–7)

Of course the Vienna Circle did not simply accept Wittgenstein's view as stated in the *Tractatus*. 'We learned much by our discussions of the book,' Carnap later wrote, 'and accepted many views *as far as we could assimilate them to our basic conceptions*' (Carnap 1963a, pp. 24–5, our emphasis). The Vienna Circle's conception of language would, inevitably, be quite different from Wittgenstein's, which was curiously detached from any actual use. It was an abstract account of language in general, completely untroubled by any actual applications. As Michael Dummett (1981, p. 679) puts it, 'The *Tractatus* is a pure essay in the theory of meaning, from which every trace of epistemological or psychological consideration has been purged as thoroughly as the house is purged of leaven before the Passover.' But it is not just unattached to any roots in sensory cognition; it also remains curiously isolated from the abstract languages of pure and applied mathematics. As a 'pure essay in the theory of meaning' it floats freely *between* the ground of sensory knowledge and the higher reaches of theoretical abstraction in science and mathematics.

So to get the *Tractatus* to do what they wanted it to do – reconcile mathematics with empiricism – the Circle had to make some modifications. They had to extend the *Tractatus* conception of language in both directions, both 'downward' to sense-data and 'upward' to mathematics. Their 'downward' extension gave the *Tractatus* view an epistemological and positivistic twist, by interpreting Wittgenstein's 'atomic sentences' as elementary observation sentences. The 'upward' extension amounted to combining the *Tractatus* with logicism, so that the empty and tautological status Wittgenstein gave logic was thereby transmitted to all of mathematics. This view was neither that of the first-generation logicists nor that of Wittgenstein; to distinguish it from these better-known conceptions, we call it 'tautologicism'. Naturally, the Vienna Circle did not distinguish their doubly extended version of the *Tractatus* view from Wittgenstein's own; to them it was a single and interlocking complex of ideas.

## 2 Two problems

Even thus extended, the Wittgensteinian conception caused problems. We focus here on two that particularly concerned the Circle. First, there was the problem of self-reference, essentially stemming from Russell, applied back to the very sentences that state or spell out the conception of language and representation (the picture theory and so on). But Russell's worries about impredicativity and the dangers of allowing a general sentence to fall within its own scope take on a new character within the picture theory. If language indeed has the isomorphic representational character claimed in the picture theory, the question inevitably arises how the 'elucidatory' sentences stating that theory (which are of course also in language), qualify as meaningful: Are they themselves pictures of facts, or are they tautological? This question much preoccupied the Circle. In one of their discussions, for instance,

> *Gödel* asked how the discussion about logical questions could be justified, as it involves the utterance not of any meaningful sentences but only of elucidations [*Erläuterungen*]. This raises the question how admissible elucidations are to be demarcated from metaphysical pseudo-sentences. (ASP, RC 081-07-11; Stadler 1997, p. 288)

This brings down to bare bones a central question facing the Vienna Circle during this period: What protected its critique of traditional philosophy from *itself*? Is the verification principle *itself* verifiable?[3]

The other problem for the Vienna Circle resulted from their own 'upward' extension of the *Tractatus*, their 'tautologicism'. By tautologicism, all of mathematics (and thus most of science) is conceived as possessing the truth-functional character of meaningful language that the picture theory gives to logic. But it didn't look to the Circle as if the logic of the *Tractatus* could be extended to allow for unbounded quantification, while still retaining the truth-functionally specified characterization of logical truth. And this left it insufficient for expressing even a fragment of actually existing science. One might call this the 'finitism problem'. Moreover, there were the problematic 'axioms' of traditional logicism: infinity, choice, and reducibility; what made these tautological?[4] A familiar objection is that whatever the Vienna Circle and others may project into it, the *Tractatus* itself does not actually raise the finitism problem (e.g. Floyd 2002). But for the Circle's application of tautologicism to scientific theories, finitism seems inescapable. By their 'downward' extension of the *Tractatus*, they took 'elementary proposition' to mean something like 'observation protocol' (Carnap's 'elementary experience', Machian 'element') – as indeed Wittgenstein himself seems to have done at least sometimes during this period; in a conversation of 1930–1 he says that 'object' in the *Tractatus* is 'used for such things as a colour, a point in visual space, etc.' (Wittgenstein 1980, p. 120). If a scientific theory is a *truth function* of observation sentences, then it *can* only be a statement about a *finite* number of instances, not a universal law, since the number of observations is always finite. This was why the picture theory, combined with the Circle's empiricism, made theoretical science as ordinarily conceived impossible.

Wittgenstein himself had of course confronted the first problem of self-reference head on, taking the heroic position that his own statements were, in the light of what they themselves asserted, strictly nonsense, and that you have to 'kick away the ladder' once you see things clearly. This is developed from the 'saying'–'showing' distinction arrived at from the impossibility of referring to the structure of language itself, in the 4.12s, and then applied with full force in the final sentences.

The Circle of course resisted this proposed solution. For one thing, it conflicted with their central project of rational reconstruction. If discourse about language is excluded, then it becomes impossible to compare different expressions. It becomes impossible to say, for instance, that a rationally reconstructed concept is *more precise*, or *more useful*, than the concept to be reconstructed. This obstructs the Vienna Circle's critique of metaphysics and unclear thinking, and undermines its entire Enlightenment project. On a more basic level, the Vienna Circle had little use for the 'pure theory of meaning' purged of any application to real life. That the basic principles of a theory should have their own meaninglessness as a consequence could only be regarded as a new and refined form of reductio ad absurdum.

But given this disconnect with Wittgenstein's agenda, you might say, did the Circle even grasp Wittgenstein's *problem*? The Russellian background of wrestling with problems of self-reference was clearly the fountainhead of Wittgenstein's characteristic doctrines. It was the very specific conjunction of these Russellian problems of self-reference and the picture theory that set the problem to which Wittgenstein responded with his own very distinctive conclusions (e.g. regarding the ineffability of logic and ethics). And these conclusions associate the *problem* they respond to with much broader philosophical issues of a kind that the Vienna Circle had turned their backs on. But this conjunction (Russellian self-reference problems plus picture theory) does not *have* to be taken in Wittgenstein's direction. It does not constitute a misunderstanding of the *problem*, just a resistance to Wittgenstein's *solution*, to take a rather 'minimalist', technical approach to the elucidation problem. This puts it back into the Russellian context where it originated, rather than branching off into the Wittgensteinian depths. So the Circle's hesitation in following Wittgenstein this far is not in itself evidence that they misunderstood Wittgenstein's *problem*, only that they did not share his goals.

## 3 What did Carnap need Wittgenstein for?

These two problems were quite serious, and for Carnap in *particular*, it's worth digging a little deeper, and asking again why, *despite* those problems, the Wittgensteinian framework was so compelling. For it turns out that in fact, he had previously developed his *own* version of a picture theory. His earliest conception (1922) of the constitution system that later became the

*Aufbau* makes the 'structural' representation of knowledge sound remarkably similar to Wittgenstein:

> *Theses*. I. The sense [*Sinn*] of every scientific statement consists in this: that a particular formal structure is ascribed to a particular piece of reality [*Wirklichkeitsstück*].
>
> II. An object within reality [*Ding der Wirklichkeit*] is identified and encompassable [*erfaßbar*] within a scientific statement only when its [conceptual] neighborhood [*Gebiet*] is put in correspondence with a constellation of a particular structure ('structural reconstruction') and it is itself put in correspondence with a particular element of this constellation. (ASP, RC 091-17-12c, p. 1v)

He acknowledges that these two theses seem mutually circular; each refers to the other. His tentative solution to the apparent circularity is to suggest a structural criterion for the whole of knowledge, in which the later *Aufbau* idea of 'purely structural description' (as exemplified in the railway map example of §14) is already evident:

> The circularity that appears to reside in the mutual reference of these two theses to each other is to be solved as follows: science, insofar as it treats of reality, initially has the task of putting every sphere of reality [*Wirklichkeitssphäre*] into correspondence with a sufficiently differentiated constellation, i.e. one in which no two members are structurally similar when the corresponding elements of reality are not identical. When that is the case for all elements of a sector of reality, then the demands of the two theses are met and thus the first task of science, the identification of its objects, achieved. (ibid.)

In Wittgenstein's way of putting it, the proposition and the pictured facts must have the same 'multiplicity' (Wittgenstein 1922, 4.032–4.0412).[5] So Carnap had his own picture theory, and was even aware of a certain circularity in it that made the picturing relation itself impossible to represent directly. What did he need Wittgenstein for? It was the picture theory, after all, that had made the logical truths (and thus, by tautologicism, mathematics) empty and tautological. What was it about Wittgenstein's *particular* form of the picture theory that Carnap needed? It seems insufficient to say that the difference between them was merely one of emphasis, for Carnap states unequivocally – even much later, in the 1950s – that apart from Russell and Frege, Wittgenstein influenced him more than anyone else.

The essential insight of Wittgenstein's that really was critical for Carnap's further development after 1926 was the idea that the truths of logic are *artefacts* of the representation system for any kind of knowledge, and that as such they are *empty* (just a by-product of representing, not *part* of what is being represented). We believe that this essential insight was specifically attractive to Carnap because it *unified* two basic components of his pre-Wittgenstein constitution system. He had begun in 1922 by fusing a phenomenological account of the system's *basis* with a logicist account of the *constitution of knowledge* on that basis (Carus 2007, ch. 5). By 1925, though, he had come to see these two components as incompatible,[6] and had sought to replace as much of the phenomenology as possible by logic alone, i.e. to obtain a *purely* structural account. But what was the source of 'structure'? Logic and empirical knowledge still seemed heterogeneous; logic was applied to 'elementary experiences' from some external vantage point, as in the vestigial Kantianism of Frege's, Russell's, Hilbert's, and Poincaré's ideas about the source of logical truth. Wittgenstein's conception overcame this heterogeneity, and eliminated all vestigial Kantianisms, leaving only *one* source of truth, the empirical source. Logic resulted as a mere by-product of representing that single kind of knowledge.

Nonetheless, for all his appreciation of Wittgenstein's conception of logic, Carnap thought that its character as free-floating 'pure theory of meaning' limited its usefulness, so that those 'upward' and 'downward' extensions were urgently necessary. By the same token, though, he was painfully aware that his own carefully worked out downward extension, in the *Aufbau*, was still vulnerable to attack on its other flank – it left itself open to precisely the same kinds of objections of over-abstraction and irrelevance to actual knowledge that he himself levelled at the *Tractatus*. Hans Reichenbach and Eino Kaila[7] had taken exception, for instance, to the apparent exclusion from the constitution system of certain modes of inference required in actual science, such as empirical induction, probability, and statistical inference.

## 4 A new foundation of logic

An even more fundamental problem for the Vienna Circle's 'tautologicism' (its 'upward' extension of the *Tractatus* system) was raised by the axiomatic systems in which much of mathematics and science is framed. The explicit definitions in which Carnap had (nominally, at least) attempted to construct the whole of knowledge in the *Aufbau* could not accommodate the 'implicit definitions' of concepts in axiomatic systems that Schlick (1925) and Einstein (1921) had given such prominence. Carnap addressed this in a large-scale project to reconcile axiomatic definitions with logicism, and to transform implicit into explicit definitions. The result was a large, unfinished manuscript entitled *Investigations in General Axiomatics*. The central theorem of this manuscript, the *Gabelbarkeitssatz*, 'proves' that an axiom system is categorical if and only if it is complete (*Entscheidungsdefinit*). Arithmetic, in particular, is therefore complete, as the Peano axioms are categorical.[8]

One important feature of the system described in the *Axiomatics* was that axiomatic systems are not purely syntactic, but are given a fixed range of interpretations within a 'basic system', a *Grunddisziplin*, as Carnap called it, of arithmetic and set theory. This made it possible to regard axiomatic systems as having a definite content, as long as it could be shown that the sentences of the *Grunddisziplin* itself had definite meanings. So not only is every sentence in the language of arithmetic determinate under this view, but it has a definite meaning as well, since it is interpreted in the *Grunddisziplin*.

But where does this *Grunddisziplin* actually come from? It was all very well to show that implicit definitions become explicit definitions *relative* to an absolute system like that. But how do you get the absolute system itself, what *makes* it absolute? Carnap did not address this in the *Axiomatics* manuscript. He did address it, though, in another manuscript he called 'Neue Grundlegung der Logik' ('New Foundation of Logic'), written while he was at Davos in April 1929, witnessing the confrontation between Heidegger and Cassirer (Friedman 2000).

What he attempts in this 'New Foundation' is to erect a Hilbertian axiomatic structure on a Wittgensteinian basis. He uses Hilbert's idea of leveraging or bootstrapping the whole of classical mathematics as a kind of formal adjunct to a concrete, finitist, secure 'meta-mathematics'. Except that in place of Hilbert's finitistic metamathematics, he uses a Wittgensteinian language of truth-functional concatenations of atomic sentences at the basis – extended to include arithmetic by tautologicism.

The 'New Foundation' is really, then, a sketch of how to *frame* tautologicism so as to solve the finitism problem. In this sketch, the atomic sentences are pictures of elementary facts, as in the *Tractatus*. A 'logic' results from the addition of further signs, connectives, that are assigned no meaning, to begin with, beyond their truth tables. Also added are inference rules (modus ponens and substitution). All sentences containing the meaningless signs still have a definite meaning, Carnap argues, as they *confine* the total space of possibilities to certain rows of the truth-table of a complete truth-functional state-description of the world (of the kind envisaged by Wittgenstein). The only requirement of a 'logic' so constructed is that it not yield any atomic sentences absent from among the premises. So the connectives drop out again in the final step of an inference, in the spirit of Wittgenstein's remark:

> The sentence of mathematics expresses no thought. In life it is never the mathematical sentence we need. We use the mathematical sentence *only* to derive sentences that do not belong to mathematics from other sentences that also do not belong to mathematics. (Wittgenstein 1922, 6.21–6.211)

Within the frame of such a logic, we can then add whatever axioms we decide we want to constitute our language. And Carnap really does mean *any* axioms whatever: first of all, there are the axioms you need to get the *Grunddisziplin*: axioms for unbounded quantifiers, for arithmetic, for set theory, and so on. But beyond that, geometrical axioms, the laws of physics, and also '*any* sort of non-empirical axioms... such as an "axiom of induction" or Kantian pure principles' (UCLA, RC Box 4, CM13, item 3, p. 1) – whatever you please. (This is why he calls the idea of the 'New Foundation' *radical* formalism – despite, *unlike* Hilbert, regarding axiom systems as having a definite content.[9])

The upshot is that all inference based on these axioms becomes tautological. In Wittgensteinian terms, the conclusion means no more than the premises. And all this axiom-based inference is now on the same level; there is no distinction in principle between logical inference, physical inference, statistical inference, or any other kind. (There is his answer to Reichenbach and Kaila!) Such axiom-based inference is to be distinguished, he says, from the mere discernment that two different truth-functions of atomic sentences happen to be extensionally equivalent. Carnap called this an *empirical* equivalence (which is merely contingent). An equivalence by virtue of the constitutive axioms (which is necessary), he called an *analytic* equivalence.

An 'analytic equivalence' is true by virtue of the axioms constituting the language. An 'empirical equivalence' just says that two expressions have the same extension, correspond to the same atomic facts, but it isn't true by virtue of a language-constituting principle. So 'All featherless bipeds are rational animals' is a merely empirical equivalence, while 'f = ma', assuming Newton's Laws are among the constituting axioms, is an analytic equivalence.

This whole episode of the 'New Foundation of Logic' is quite interesting. Although it is still articulated within the picture-theory framework, this is where the idea of introducing principles of science as language-constitutive (like the 'P-rules' of the *Logical Syntax*) and, more generally, the idea of 'analytic truth' as *constitutive* of a language seems to originate. That is obviously important for the future.

## 5 Wittgenstein's prison

For the present, though, things did not look so good. In the course of 1930, the somewhat shaky 'New Foundation of Logic' collapsed. Three developments undermined it, all stemming from the earth-shaking work in mathematical logic during these years. First, the *Gabelbarkeitssatz* fell victim to Gödel's first incompleteness theorem. As Gödel indicated in the discussion following the famous symposium on the philosophy of mathematics in Königsberg in September 1930 (at which Carnap had been the spokesperson for logicism, Heyting for intuitionism, and von Neumann for formalism), there could be true arithmetic sentences that were not provable:

> One can even (given the consistency of classical mathematics) give examples of sentences (of the kind stated by *Goldbach* or *Fermat*) that are correct in their content, but not provable in the formal system of classical mathematics. By adding the negation of such a sentence to the axioms of classical mathematics, one obtains a consistent system in which a sentence whose content is false is provable. (Hahn et al. 1931, p. 148)

So despite the categoricity of the Peano axioms, arithmetic was not complete, and the *Gabelbarkeitssatz* was false. The *Axiomatics* project, Carnap's attempt to reduce implicit definitions to explicit definitions, had failed.

Second, the incompleteness result had an even more fundamentally devastating effect on logicism itself, which the Vienna Circle had relied on to guarantee the tautological (and thus empty) character of mathematics. The Circle had needed this to undermine the fundamental tenet of metaphysics that conclusions about the real world could be reached by reasoning alone, without factual knowledge (Carnap 1930a, p. 25). But now it turned out that there could be sentences of arithmetic that, despite the logicist construction of the numbers, were not logically determinate after all. The Vienna Circle's 'tautologicism' had failed.

Third and more generally, the new work in mathematical logic, especially by Hilbert, Gödel, and Tarski, was fundamentally incompatible with the picture theory. This work made essential use of the distinction between a language and its metalanguage. Through much of 1930, Carnap still clung to the forlorn hope that a single-language system along the lines of the *Axiomatics* could still be worked out, showing that although the use of a metalanguage might be practically necessary, it was not theoretically indispensable. But by the end of the year, especially in the light of Gödel's incompleteness proofs, it seems clear that the new metamathematical work represented a clear counterexample to Wittgenstein's doctrine that sentences about the structure of language are not expressible in language.

The efforts to solve the finitism problem and the elucidation problem within the picture theory had not only failed, then, but the very framework that had solved the problem of how to reconcile mathematics with empiricism – the picture-theory framework – was now turning out to be at odds with the mathematical approach that had caused the solutions to fail.

This, then, was the situation at the end of 1930: Wittgenstein's revolutionary insight, which the Vienna Circle had relied on to reconcile mathematics with empiricism, had been to recognize that the laws of logic are not laws of anything out in the world, but laws of representation. Symbolic representation of any kind has the laws of logic as its inescapable artefact. In Kantian terms, the possibility of representation determined the forms of logical intuition. The logic built into any possible language, any possible representational system was, like a Kantian form of intuition, an inescapable straitjacket. The very nature of language, in Wittgenstein's view, prevented us from ever stepping outside it. One could call this conception 'Wittgenstein's prison'. In the mid-1920s, it hadn't looked quite so much like a prison to the Vienna Circle. On the contrary, Wittgenstein had given them a great sense of liberation – from the traditional empiricist (Millian or Machian) conception of mathematics. But the further they went in attempting to extend Wittgenstein's logic upward and downward, the harder they tried to solve the elucidation and finitism problems, the more they felt trapped.

Or rather, the more *some* of them felt trapped. The 'right wing' of Schlick and Waismann did not. They accepted the inevitability of confinement and sought to make themselves at home in their prison. But the 'left wing' looked desperately for an exit. They hoped against hope that it might turn out not to be a prison after all, and some door somewhere might have been left ajar. But by the end of 1930, the exits that had previously looked possible had all slammed shut. It was time to think about planning an escape.

## 6 The 'Attempt': the germ of the *Logical Syntax*

We are now in a position to understand the significance of the 'Versuch einer Metalogik' ('Attempt at a Metalogic', henceforth 'Attempt'), the 44-page shorthand document where Carnap jotted down his ideas from the sleepless night in January 1931. What we find in it is a perspective that is radically different from the Wittgensteinian one of the 'New Foundation'. Carnap has here adopted the fully formal, 'metalogical' point of view of Gödel and Tarski, according to which the logical language is a system of uninterpreted marks rather than meaningful signs. As he would put it in another manuscript a few weeks later:

> In the calculus we 'calculate' with the signs, i.e. we carry out operations on the sign complexes according to certain rules, without regard to the *meaning* of the signs. We *can* certainly attend to the meanings of the signs, e.g. to a certain sentence in connection with 'p', or to the meaning of the word 'or' in connection with 'v', etc. The essential thing is that in the operation rules [*Operationsvorschriften*] these meanings are not mentioned. (UCLA, RC Box 3, CM10, item 8, p. 18)

In the perspective of the 'New Foundation', the atomic sentences had been pictures of atomic facts, which gave them their meaning. In the 'Attempt', an atomic sentence is a finite sequence of superscript dots, followed by the letter '*f*' with a finite sequence of subscript dots, followed by a left parenthesis, followed by the letter '*a*' with a finite sequence of subscript dots, followed by a right parenthesis, e.g.:

$$^{\cdot\cdot\cdot\cdot\cdot\cdot}f_{\cdot\cdot\cdot\cdot}(a_{\cdot\cdot\cdot})$$

An atomic sentence was thus a certain finite string consisting of instances of finitely many basic marks [*Zeichen*] – the instances themselves being *physical* marks, having a particular location on the blackboard or on a page. These *Zeichen* of the calculus are devoid of meaning, and are treated just like the figures in pure mathematical geometry.

In the 'New Foundation', a sentence is a tautology because of what it says, or does not say, about the world. In the 'Attempt', being a tautology is a property of a string of marks that is defined entirely in terms of its outer form – the type and order of the marks occurring in it. No use is made of the 'meaning', 'designation', etc. of the marks (*Zeichen*) in defining the central notions of truth-value assignment, consequence, tautology, and the like. Carnap even mentions that the undefined notion 'true' might be better avoided entirely.[10]

From the viewpoint of modern logic, this idea may not seem particularly momentous. Even at the time, it represented no technical innovation; Hilbert and others had been treating axiomatic systems formally for decades, and the methods of Gödel and Tarski did essentially that. But although Carnap's first attempt to formulate his 'metalogic' was in terms of a particular formal system, his aim was not merely the mathematical study of one such system. His new idea was to *apply* the insights of Hilbert, Gödel, and Tarski to the entirety of scientific knowledge. As we saw above, he had previously accepted Wittgenstein's basic account of the logical language framework in which all science was to be expressed, as the basis for the project of rational reconstruction. In that context, the new 'metalogical' perspective of regarding language purely as a system of formal rules, without reference to anything outside itself, was indeed a revolutionary idea.

Before Wittgenstein, language had been regarded as an essentially transparent medium for the expression of thought. The laws of logic were considered by Frege and Russell to be laws of thought, judgement, or perhaps nature – but certainly not of language. Wittgenstein had recognized that they were laws of language. But he had arrived at this idea via a theory of representation that forced language to conform always and everywhere to particular laws, arising necessarily from the representational function of language – the picture theory. The possibility of representation determined a particular form of linguistic intuition, so to speak. This basic logic built into our form of representation was, like a Kantian form of intuition, an inescapable straitjacket. The very nature of language, in Wittgenstein's view (at least as seen by the Vienna Circle), prevented us from stepping outside of it. In section 5 above, we called this quasi-Kantian view 'Wittgenstein's prison'.

Under the suggestive influence of Hilbert's formal approach to axiomatic systems and its use by Gödel and Tarski, Carnap was able to escape from Wittgenstein's prison by taking Wittgenstein's own idea of language as governed by a system of rules one step further. Carnap distinguished the representational or meaning function of language from its purely combinatorial one, and now took the *latter*, rather than the former, as his starting point. Ironically, the metalogical methods developed in pursuit of the very mathematical results (such as the incompleteness theorem) that had undermined his Wittgensteinian position in the 'New Foundation', also showed a way of breaking out of Wittgenstein's prison, and making the structure of language itself the object of logical study. As opposed to the confinement of all possible knowledge within the absolute constraints imposed by the (naturally or metaphysically) fixed structure of our means of expression, this new recognition – that linguistic structure could itself be investigated – opened up a whole new method for the unification and clarification of knowledge. Thus Carnap retained Wittgenstein's language-dependence of knowledge, but threw off the shackles of Wittgenstein's prison in favour of the logicians' metalogical perspective.

Armed with this new insight then, and in the rush of enthusiasm that accompanied it, Carnap apparently hoped to be able to solve the other problems that had undermined the 'New Foundation', particularly those afflicting logicism. Arithmetic, it was envisaged in the 'Attempt', could evidently somehow be 'read off' from the syntax of the logical object language – as opposed to being expressed in that language.[11] Thus the numbers are there not defined as higher-order concepts in the Frege–Russell logicist style, but 'purely as figures' (*rein figurell*), on the basis of the dot sequences attached to the symbols. Arithmetical properties and statements then belong to the metalanguage. Thus e.g. the commutativity of addition $n + m = m + n$ was supposed to follow from the fact that n-many dots written to the left of m-many dots gives the same series of dots as writing them to the right of m-many dots. The question of the need for mathematical induction *in the metalanguage* is briefly considered, but then dismissed with some optimism.

If arithmetic was to be formulated in the metalanguage of logic, then analysis was to be formulated in its meta-metalanguage. For real numbers are properties or series of natural numbers, and properties of them and statements about them properly belong one level up. Carnap may have been guided, in this idea, by Russell's suggestion, in his introduction to the *Tractatus*, that one could perhaps break out of Wittgenstein's prison by using a scheme involving a hierarchy of languages:

> These difficulties suggest to my mind some such possibility as this: that every language has, as Mr. Wittgenstein says, a structure concerning which, *in the language*, nothing can be said, but that there may be another language dealing with the structure of the first language, and having itself a new structure, and that to this hierarchy of languages there may be no limit. (Russell 1922/1988, p. 286)

Having now found the mechanism for such a scheme in the form of 'metalogic', applying it to achieve a hierarchy consisting of language, metalanguage, meta-metalanguage, and so on[12] must have indeed seemed rather compelling, at first sight.

Carnap says in his autobiographical account that not only 'the whole theory of language structure' came to him like a vision, but also 'its possible applications in philosophy'.[13] These were spelled out later that year in the paper 'Die physikalische Sprache als Universalsprache der Wissenschaft', which was later published in English, with a new preface, as the pamphlet *The Unity of Science*. This paper is mainly known for its advocacy of physicalism, and is thus taken to represent a watershed in Carnap's epistemological views from the phenomenalism of the *Aufbau* to a Neurath-inspired physicalism. This epistemological aspect is certainly present in the paper, and reflected in its title. But the new syntactical doctrine is equally in evidence and, indeed, motivates the paper's physicalistic conclusions. After three pages of introductory discussion about the idea that all objects and facts are of a single kind, we are told that these expressions are a concession to the customary 'material' (*inhaltliche*) way of speaking. The 'correct' way, Carnap says, speaks of words rather than 'objects' and sentences rather than 'facts', for a philosophical investigation is an analysis of *language*. In a footnote he indicates that a comprehensive, strictly formal theory of language forms, which he calls 'metalogic', will soon be forthcoming, and will justify the 'thesis of metalogic' here invoked, that 'meaningful' (*sinnvolle*) philosophical sentences are the metalogical ones, i.e. those that speak *only* of the form of language (Carnap 1932b, p. 435).

This represents a radically different basis for the critique of metaphysics from the one Carnap had previously adopted from Wittgenstein, whereby meaningful sentences were those that derived their meaning from atomic sentences by truth-functional combinations. Atomic sentences, as pictures of atomic facts, no longer play any role in distinguishing meaningful from meaningless sentences. The new metalogical or syntactic viewpoint is significant, as Eino Kaila agreed after discussion with Carnap a few months later, because of its 'elimination of verification by comparison with facts [*Ausschaltung der Verifikation durch Vergleich mit Sachverhalten*]' (ASP, RC 025-73-05: diary entry of 26 June 1931).[14]

So Carnap had comprehensively and definitively turned his back on the picture theory of the *Tractatus* – and thus also on its foundationalism. Meaning was no longer built up from some basic (naturally occurring or metaphysically unavoidable) components. The rules of syntax were no longer to be *discovered*, for they were no longer objectively determined by the mechanism of representation, as Wittgenstein had seemed to suggest. Instead, they were a matter of human decision, conventions by which we set up the language of science. And the resulting possibility of 'engineering' the logical language in the service of science became the source of the exciting 'possible applications in philosophy'.

In maintaining Wittgenstein's language-dependence of knowledge, but casting off the necessary form of the structure of language as a constraint on our knowledge, Carnap thus finally arrived at a basis for the Vienna Circle's original Enlightenment programme; it became the programme of unifying and advancing knowledge through logical engineering of language. The upshot of Carnap's dream, then, was a liberation from the manacles of a fixed structure imposed on the mind by natural or metaphysical factors beyond our control, and the recognition of its potential as a vehicle for human improvement. January 1931 was thus the turning point in Carnap's philosophical development; with respect to Wittgenstein's prison, he went from slave to master literally overnight.

## 7 The 'Metalogic': the first draft of *Logical Syntax*

In rejecting Wittgenstein's doctrine of nonsense in favour of the logicians' conception of metalogic, the first of the two problems discussed in section 2 above, the problem of self-reference, was solved: philosophical elucidations could be regarded as perfectly legitimate, meaningful *metalinguistic* statements about the scientific *object language*. Whether these statements, in turn, conformed to *meta-metalinguistic* analogues of their own strictures was another, perfectly legitimate, question that could be straightforwardly considered; there was no contradiction involved in such questions, and no need for the philosophical acrobatics involved in Wittgenstein's kicking away of the ladder.

But the excited solution to the problems of logicism suggested by the new metalogical standpoint turned out not to work. The rather odd idea that arithmetic could be read off from the metalanguage of logic in a sense turned out to be *too* correct, in that some essential metalogical concepts (notably *provability*) required for their formulation a combinatorial theory that was every bit as complicated as arithmetic itself. Thus in the late spring of 1931, Carnap decided to move to a conventional axiomatic arithmetic in the *object* language, so that the axiomatized arithmetic could then be used to express the metalanguage, using Gödel's method of arithmetization (ibid., §19). This move had the further advantage of collapsing the entire hierarchy of languages and metalanguages into itself, at least in principle, by iterating Gödel's method of arithmetizing the metalanguage in the object language. Thus it appeared (for a time at least) that one could now get by with only a single language after all.[15]

However well this seemed to work, there was still a price to be paid for it. For the very thing that had made the 'metalogical' solution possible in the first place – i.e. the precise definability of the central metalogical notions and their resulting expressibility in the object language – was also responsible for the essential incompleteness of the logical treatment of mathematics. The identification of the logical with the formal thus seemed to restrict its scope to only what can be captured with limited means. An idea for a solution to this central problem seems to have occurred to Carnap quite early on in the development of the *Syntax* programme, perhaps influenced by his conversations with Tarski in 1930: if there were no intrinsic constraints of 'correctness' on the sorts of formal properties of formulas that could be considered, then perhaps there could be a formal criterion for mathematical truth *different* from mere provability. Since Gödel had shown that provability was insufficient – there were 'true' arithmetical statements not derivable from the axioms – the identification of such a criterion was essential. Carnap apparently developed such a criterion sometime in the latter part of 1931, in the form of the notion of *analyticity*. This was to be a stronger sort of logical truth than provability in a formal system, but was still to be determined strictly in terms of the formal character of the symbols.

Analyticity was apparently to take the place of provability as the generalized notion of tautology or logical truth for the purposes of tautologicism. To understand how this was intended, consider the analogy of a chess game. Think of the starting position of the pieces as the axioms, the permitted moves as the rules of inference, and a sequence of moves ending in checkmate as a proof of a theorem. But now observe that there are configurations of pieces on the board that constitute checkmate, but cannot be reached from the starting position by any sequence of permitted rules. Such a configuration represents an analytic sentence that has no proof. In this way, the definition of analytic sentence can be phrased entirely formally, in accordance with all the same rules of inference, and yet still be wider than provability. Thus the absolute, Wittgensteinian conception of tautology could be saved, and indeed finally extended beyond propositional logic in accordance with the Vienna Circle's original ambitions.

Such a notion of analyticity was apparently defined in the first draft of the *Logical Syntax*, entitled 'Metalogik', of which nothing has been preserved (as far as we have been able to determine) but its table of contents. This lists the notion *analytic* alongside *synthetic* and *contradictory* under the heading 'IV.B. Theory of content of formulas' (corresponding roughly to IV.B(a) of *Logical Syntax*, which – in the English translation – gives the general definition of 'analytic'). This is followed in section IV.C by a discussion of soundness, consistency, and completeness, including sections on the 'antinomies' and 'the incompleteness of all formal systems' which appear to correspond closely to IV.B(c) of the (English) *Logical Syntax*, where the Gödel incompleteness of arithmetic is discussed.

We don't know exactly how analyticity was originally defined, but from the evidence available it is clear that the definition was defective. As we shall explain presently in more detail, Gödel objected to its application to the 'extended model language'.[16] And furthermore, he points out, it will be *impossible* to give a correct definition of it in *any* metalanguage that can be faithfully represented in the object language, e.g. by arithmetization. This fact has since become known as Tarski's theorem on the indefinability of truth. Thus it turns out that Carnap's single language approach will not work after all.

Gödel's objection to Carnap's original definition of analyticity is explained in a letter dated 11 September 1932 (Gödel 2003, pp. 346–8). Carnap had apparently tried to define the notion 'analytic sentence' inductively, using what we would now call a substitutional treatment of quantification. Thus e.g. given an arithmetical sentence of the form $(\forall x)\, f(x)$, with quantification over the numerical variable $x$ and $f(x)$ a formula with at most $x$ free, one could reasonably define:

$$(\forall x) f(x) \text{ is analytic } \Leftrightarrow_{\mathrm{df}} \quad f(a) \text{ is analytic for all numerical expressions } a$$

In his definition, Carnap had apparently tried to use the same strategy for higher-order quantifiers, for example over all properties or sets, as in $(\forall X)$ $f(X)$. Thus e.g. for $f(X)$ of the simple form $X(0)$ one would have:

$$(\forall X) X(0) \text{ is analytic } \Leftrightarrow_{\mathrm{df}} \quad A(0) \text{ is analytic for all predicate expressions } A(x)$$

But here there is no restriction on what predicate expressions (i.e. 'open sentences') $A(x)$ are to be substituted for the variable $X$ in testing for analyticity, so among the substitution instances is e.g. the predicate $(\forall X)\ X(x)$ itself. Thus the definition is circular, and so it does not succeed in specifying the desired notion. The problem here is in the so-called 'impredicativity' of the higher-order quantifier. One could restrict the substitutions to predicates of lower 'order', in a suitably defined sense, and this would result in a workable scheme, but it would only provide a definition for a system like ramified type theory, which is inadequate for classical mathematics.

In his letter, Gödel suggests instead using a notion of 'all sets and relations whatever' (*alle Mengen und Relationen überhaupt*) in place of 'all predicates'. An interesting footnote indicates that this need not be interpreted as Platonism, as he only suggests formulating the definition of 'analytic' in a particular metalanguage, in which the concepts of 'set' and 'relation' are already given. He goes on to say that he intends to use this idea to give a truth definition in Part II of his paper (presumably the missing sequel to Gödel 1931). And, moreover, that he believes it cannot be done otherwise, and that the higher functional calculus can*not* be treated 'semantically' (i.e. according to Carnap's strictly formal conception of metalogic).

In his first reply, a desperate Carnap attempts to reconstruct Gödel's proposal – the difficulty lies in the idea of 'all values' for a predicate of the object language *L*. How is this to be understood, even with respect to another language *L′* in which the values are to be taken? It will not suffice to use only the predicates definable in *L′*; one apparently needs instead all 'arbitrary' ones. And this latter notion strikes him as rather questionable (*ziemlich bedenklich*). He finally asks for help in finding the right definition, especially since, as he says, everything else in his book depends on it (ibid., pp. 350–2).

Judging from his note of a few days later, Carnap finally did work out the solution for himself. He realized that the notion of 'all values' of a predicate could be rendered in the formal meta-language *L′* simply by using a universal quantifier $(\forall X)\ \ldots X \ldots$. The key new idea here is that the language *L′* in which the values are taken needs to be stronger than the one for which they are given (ibid., p. 354). In his (delayed) reply, Gödel confirms that this is the idea, and remarks that one cannot give the definition of 'analytic' in the same language, otherwise 'contradictions will result'. He also points out that, presumably in the meantime, Tarski has already published a 'similar' definition of 'analytic',[17] which seems likely to be the reason Gödel never worked out his own part II (ibid., p. 356).

For Carnap, ultimately, the resulting definition of 'analytic' – which had previously been so crucial – was not even deemed important enough to include in the first edition of the book; it was omitted 'for reasons of space'.[18] The problem with it was that, as hinted by Gödel in the footnote about Platonism, the notion of analyticity it defined was not absolute, but rather in a certain sense, conventional. It gave a notion of 'analytic in *L*', but only with respect to *another* language *L′*, used for the interpretation of *L*. There might be a natural or conventional choice for *L′* – type theory of the next higher type, or axiomatic set theory – but it could hardly be claimed that any particular such choice is the *correct* notion of analytic for a given language. This *language relativity* of the central notions of metalogic turned out to be more important to Carnap than the particular metalogical definitions themselves. And this brings us to the other step in Carnap's creation of the *Logical Syntax*.

## 8 The 'principle of tolerance'

> The view that the terms 'analytic' and 'contradictory' are purely formal and that analytic sentences are empty of content was stated by Weyl … Later, Wittgenstein made the same view the basis of his whole philosophy. 'It is the characteristic mark of logical sentences that one can perceive from the symbol alone that they are true; and this fact contains in itself the whole philosophy of logic.'… 'And so also it is one of the most important facts that the truth or falshood of non-logical sentences can *not* be recognized from the sentences alone.' This statement, which gives expression to Wittgenstein's absolutist conception of language and leaves out

> the conventional factor in the construction of a language, misses the mark. (Carnap 1934d, p. 139)

The first public signal that Carnap's thought had entered yet another radically new phase was the discussion contribution 'Über Protokollsätze'. It was written within a month or two after the above correspondence with Gödel, and Carnap is a changed man. A new tone has suddenly entered his writing, one that would become deeply characteristic: 'In my view the issue here is not between two conceptions that contradict each other, but rather between *two methods for constructing the language of science, which are both possible and justified*' (Carnap 1932c, p. 215). And he spells out the grounds of this new pluralism:

> Not only the question whether the protocol sentences are inside or outside the syntax language, but also the further question regarding their precise specification, is to be answered, it seems to me, not by an assertion, but by a stipulation [*Festseztung*]. Though I earlier [in 'Die physikalische Sprache'] left this question open . . . I now think that the different answers are not contradictory. They are to be taken as proposals for stipulations [*Vorschläge zu Festsetzungen*]; the task is to investigate these different possible stipulations as to their consequences and assess their usefulness. (ibid., p. 216)

We have no record of the moment at which Carnap embarked on this new direction, but the sense of discovery and enthusiasm is palpable in 'Über Protokollsätze'; he repeats the new message again and again. And he is very much aware that it represents an even more radical departure from his and the Vienna Circle's previous position:

> In all theories of knowledge to date there is a certain *absolutism*: in the realistic theories an absolutism of objects, in the idealistic ones (including phenomenology) an absolutism of the 'given', of 'experiences', of 'immediate phenomena [*unmittelbare Phänomene*]'. Even in positivism we find this residual idealistic absolutism; in the logical positivism of our circle – in the works on the logic of science (epistemology) published to date by Wittgenstein, Schlick, Carnap – it takes the more subtle form of an absolutism of primitive propositions ('elementary propositions', 'atomic propositions'). (ibid., p. 228)

This sense of breakthrough is equally evident in the passages evincing this new 'principle of tolerance' in the *Logical Syntax* itself. In the messianic preface he writes:

> The range of possible language forms, and thus of different possible logical systems is ... incomparably larger than the very narrow range in which modern logical investigations have so far operated. Up to now there have only been occasional small departures from the language form given by *Russell*, which has already become classical ... The reason for not daring to depart further from this classical form would appear to lie in the widespread view that such departures must be 'justified', i.e. it must be shown that the new language form is 'correct', that it represents the 'true logic'. It is one of the main tasks of this book to eliminate this view as well as the pseudoproblems and pointless squabbles arising from it. (Carnap 1934d, p. v)

Instead, he urges the view that we have total freedom in choosing the form of language, and he concludes the preface with the colourful prose:

> Those early attempts to cast the ship of logic off from the solid shore of its classical forms were surely daring, historically speaking. But they were hindered by a striving for 'correctness'. Now this barrier is overcome: before us lies the boundless ocean of free possibilities. (ibid., p. vi)

The new perspective first appears in the text of the *Logical Syntax* itself in the context of philosophies of mathematics, particularly intuitionism. There it occurs as the exhortation to state metatheoretic or *wissenschaftslogische* proposals in precise terms, as explicit rules or definitions, within the formation or transformation rules of a precisely defined language or calculus:

> Once it is understood that all pro- and anti-intuitionist considerations are concerned with the form of a calculus, the question will no longer be asked in the form 'What *is* the case?' but rather 'How do we *want* to set this up in the language being constructed?' ... And with that, the dogmatic frame of mind that often makes the discussion unfruitful is banished. (ibid., p. 42)

This 'dogmatic frame of mind' results, in Carnap's view, from the reliance on inherently vague philosophical 'considerations' (*Erörterungen*) rather than on precise statements of definitions and rules. Only by *replacing* the vague concept with a precise equivalent can the practical merits or drawbacks of a proposal be judged, for some defined purpose. And under the new regime of pluralism, where there can be no criterion of inherent 'correctness', practical usefulness is the only criterion left for deciding whether a proposal should be pursued or left aside.

The principle of tolerance thus fits well into the project of 'rational reconstruction' pursued by the earlier Vienna Circle (and sets the stage for the successor project of 'explication', which Carnap would not formulate explicitly until after 1945).[19] It also resonates with the conventionalism of Poincaré, Duhem, and Einstein, which had always been an important strand in the Circle's brand of positivism. But now that conventionalism was being extended to logic itself, by applying Gödel's metalogical standpoint to the Wittgensteinian recognition of the essential role of the syntax of language in determining logical truth.

Unlike some previous revolutions in Carnap's thought, this one was permanent; it became the basis of his thought for the remainder of his career (Carus 2007, ch. 10). The new conception was only partially spelled out in the *Logical Syntax* itself, and it was only indicated generally in such later writings as 'Empiricism, Semantics, and Ontology' and certain replies to critics in the Schilpp volume.[20] But it extended, as we have seen, far beyond logic itself into epistemology and such varied questions as the form of the scientific observation language and the purpose of philosophy. It represents the second and final step away from the meaning foundationalism of the *Tractatus*, to a kind of radical pragmatism, in which the only criterion for acceptance or rejection of a language form is its usefulness for a particular purpose. And as we shall argue below, this mature position was intimately tied to some of the principal themes and controversies in twentieth-century philosophy. Carnap's sleepless night, in which he was first visited by the dream of the *Logical Syntax*, thus not only marked a turning point in his own philosophical development, but – like the dreams of Descartes – in the development of philosophy.

## 9 Two applications

We conclude by considering two different applications of our two-step analysis of the genesis of the *Logical Syntax* to subsequent philosophical discussions.

We first ask to what extent the breakthrough of the sleepless night, the original syntax idea, was conceptually necessary for the subsequent move to tolerance. To what extent are the two steps even separable? This question is of particular interest in the context of present discussions of the *Logical Syntax*,[21] in which it is sometimes argued that tolerance depends heavily on the syntax idea, and that the later abandonment of the original syntax view significantly restricted the principle of tolerance, or made its limitations evident.[22]

Our analysis provides an answer to this question. We saw in section 6 that the original syntax idea represented, above all, a rejection of *meaning* in Wittgenstein's sense. According to the 'absolutist' view of meaning, as Carnap later called it, the meanings of all sentences rest on the representation of atomic facts by atomic sentences. One could call this view 'meaning foundationalism'. We interpreted the new view Carnap arrived at during his sleepless night in January 1931 as the replacement of this meaning foundationalism by an axiomatic approach to language as a whole, in which all workings of the language are exhaustively specified by explicit rules stated in a metalanguage. In its original statement, this 'syntax' view completely excluded the possibility of 'meaning' – in its old sense of representational correspondence between configurations of linguistic objects and configurations of objects in the world. There seemed to be no way of capturing any such correspondence in explicit formation or transformation rules for a language. But just a year after the *Syntax* book was published, Tarski's definition of truth suggested to Carnap that such correspondences could, after all, be captured in metalinguistic rules. This amounted to defining a new notion of meaning 'from above', in contrast to the one built up 'from below' in the meaning foundationalism that had been rejected. In this new scheme, the fact that the language itself was constituted by a system of rules permitted the rigorous specification of an 'interpretation' (by induction), as opposed to regarding the rules as being descriptive and determined by a more fundamental notion of meaning built up from the atomic components. Thus,

$$p \,\&\, q \text{ is true iff } p \text{ is true and } q \text{ is true}$$

is (part of) the definition of 'truth', not the meaning of '&'.

In January 1931, the rejection of meaning foundationalism and its replacement by an axiomatic view were all of a piece. But seen from the later, semantical perspective, this original syntax view could be regarded, retrospectively, as having been composed of a number of different elements that would turn out to be separable. First, there was (a) the requirement that a language be entirely specified by explicit rules. The 'syntactic' view that seemed to follow from this can in retrospect be seen to have consisted of two separable parts: (b) the distinction between a language (a calculus, a purely syntactic symbol system) and its interpretation, and (c) the prohibition of reference to the latter, and the restriction of the (*wissenschaftslogische*) metalanguage to consideration only of the former.

Components (a) and (b) are necessary pre-conditions for the tolerance idea. Without the requirement that language be specified by explicit rules, the alternatives that are to be tolerated are not fully specified. And without distinguishing language from content, there is no possibility of distinct alternatives among which to be tolerant. These two retrospectively visible components survive unscathed and undiminished into Carnap's semantic period. (So it is perhaps misleading to call them 'syntactic'; Carnap's original term 'metalogical' might be more appropriate.) Component (c), on the other hand, was an over-reaction against Wittgensteinian 'meaning' that accompanied the original insight, and did not survive. In distinguishing between a language and its interpretation, Carnap's first (and, as we saw, understandable) response was to reject that imprecise notion of meaning entirely. But this restriction was relaxed when he saw that, in virtue of the precise specification of the object language, interpretations could *also* be specified by explicit rules (governing satisfaction, designation, and truth), in accordance with component (a) of the original syntax idea. This answers the question about the extent to which tolerance depends on a specifically 'syntactic' approach. Of the components of the original syntax idea, the two 'metalogical' ones – (a) and (b) – are consistent with tolerance, while the rejection of meaning – component (c) – is inconsistent with tolerance.

The original rejection of 'meaning' had proscribed what seemed an occult property, just like the rejection by Lavoisier of the traditional explanation of burning as the release of a substance ('phlogiston', in Stahl's theory) into the surrounding air. The reinstatement of an explicated account of 'meaning' reflected the realization that the informal idea of meaning had not itself been the culprit, but rather a particular, somewhat obscure conception of it (Wittgenstein's, in the Circle's view). But the new explication of meaning met the standards by which the previous conception had been rejected. In the same way, the later reinstatement of the idea that burning (oxidation) involved the release of electrons by the substance being oxidized met the standards of the post-Lavoisier principle of the conservation of matter, by which all reactions are regarded as recombinations of indestructible elementary particles. The new explication of the informal concept of 'meaning' has no more in common with the previous occult property than electrons do with phlogiston.

While the original syntax idea was a necessary precondition for tolerance, then, one component of it (and, in the book, the most visibly high-profile component) would soon be jettisoned. This gave the appearance that the 'syntax' doctrine – identified with the exclusion of meaning – had first been embraced, and then rejected again. 'Meaning' was proscribed in 1931, it seemed, and then became acceptable again in 1935. What most observers failed to notice in this sequence of events was the permanence of the 'metalogical' (or 'top-down') components and the principle of tolerance consequent upon them. The January 1931 rejection of meaning foundationalism – which is nothing but Quine's second 'dogma of empiricism' – was permanent; as we saw above, the *main point* of the original syntax idea had been the '*elimination* of verification by comparison with facts'.[23]

This brings us to the second application of our analysis: we claim that Quine's celebrated 'two dogmas of empiricism' were not only not held by Carnap at the time the paper was published, but in fact had been definitively rejected by him some twenty years before. That the real target of Quine's critique was not the mature Carnap of the 1950s, but the younger one Quine had been overawed by in the early 1930s,[24] is evident right in the first paragraph of 'Two Dogmas', where the (second) dogma of Reductionism is described as 'the belief that each meaningful statement is equivalent to some logical construct upon terms which refer to immediate experience' (ibid. p. 20). Quine's extended critique of 'meaning' in section V amounts, then, more specifically to a critique of the 'verification theory of meaning', according to which 'the meaning of a statement is the method of empirically confirming or infirming it' (ibid., p. 37). He thus insinuates that Carnap's conception of 'meaning' circa 1950 is still essentially wed to the foundationalism of the *Tractatus*. But we have just seen that it was precisely this absolute, reductionist conception of 'meaning' that Carnap left definitively *behind* after the sleepless night on 21 January 1931. The rejection of Quine's second dogma was the substance, we might say, of Carnap's Dream. And we saw that the post-1935 semantic reconstruction of meaning has no more in common with this earlier conception than electrons have with phlogiston.

It is the first dogma, however, which has occupied more of the limelight, and Quine's terminology is not so obviously anachronistic in this case. Also, Carnap has been almost universally portrayed in the literature as defending the dogma Quine attacks. But our analysis above shows that this is simply not the case. What Quine demands of Carnap in his celebrated paper is an account of 'analyticity' in the sense of a supposed 'fundamental cleavage between truths which are *analytic*, or grounded in meanings independently of matters of fact, and those which are *synthetic*, or grounded in fact' (p. 20). More specifically, he wants something that tells us what 'analyticity' is not just relative to some particular language $L_0$ but *in general*:

> The notion of analyticity about which we are worrying is a purported relation between statements and languages: a statement *S* is said to be *analytic for* a language *L*, and the problem is to make sense of this relation generally, that is, for variable '*S*' and '*L*'... By saying what statements are analytic for $L_0$ we explain 'analytic-for-$L_0$' but not 'analytic', not 'analytic for'. We do not begin to explain the idiom '*S* is analytic for *L*' with variable '*S*' and '*L*'... (ibid., pp. 33–4)

This passage reveals how seriously Quine underestimated the radicalism of Carnap's 1931–2 deflation of Wittgensteinian 'truth' and 'meaning', which left behind the idea of an absolute or language-transcendent notion of analyticity. Indeed, this is exactly what was overcome by Carnap in the second step of the development that we have just described. As we saw in section 7 above, it was precisely the impossibility of a canonical concept of analyticity of the kind he had attempted in the first draft of the *Syntax* – pointed out by Gödel in October 1932 – that motivated the move to the principle of tolerance. Analyticity had become entirely language-relative and stipulative. There was no longer a trans-linguistic, absolute notion of analyticity; nor was there, for a given language, a distinguished choice of metalanguage in which to formulate a canonical notion of analytic truth. The logical conventionalism that resulted was precisely the second of our two steps, encapsulated in the principle of tolerance. Quine's failure to grasp the import of this step is evident in his jibe that a merely conventional definition of 'analytic-for-$L_0$' is so uninformative that it 'might be better written untendentiously as "*K*" so as not to seem to throw light on the interesting word "analytic"' (p. 33). In fact, what Quine satirizes as trivial is precisely the deflated notion of 'analytic' that Carnap employs, after 1932; he has left behind any ambition of throwing light on the '*interesting*' word 'analytic' in the sense Quine demands.

Quine himself seems not to have been entirely certain that the later Carnap actually held the views he attributed to him and criticized.[25] He even seems on some occasions to have conceded that the differences between himself and Carnap were not of a 'cognitive' nature.[26] But these caveats did not prevent the philosophical profession from regarding Quine as having been 'right' and Carnap 'wrong', nor from attributing to Carnap the views Quine criticized. Half a century is enough; it is time now to retreat from these misattributions. In the light of the story we have told above, it should henceforth be evident that Quine's supposed 'two dogmas' were exactly the two components of the 'absolutist' conception that Carnap successively overcame in creating the *Logical Syntax*: escaping from 'Wittgenstein's prison' in the night of 21 January 1931; and, with Gödel's help, finally reaching the 'boundless ocean' of possibilities in October 1932, never to look back.

## Notes

1. Our first effort to address this puzzle is our paper Awodey and Carus (2007).
2. Which now takes the updated form of claiming that the *Tractatus* has been wrongly interpreted because it has been mistakenly *assimilated* to Carnap's different conception. James Conant, for instance, attributes to early Wittgenstein the idea that meaning is in the mind of the speaker or the writer, without whom the mere physical sign is not a symbol; the meaning relation (putting the symbol into the sign) is established only in the minds of human interlocutors (Conant 2001, pp. 24–8). Whether a sign or a sentence has meaning cannot be objectively determined, according to this view, i.e. cannot be determined outside the context of its use and the intentions of its users. This view of Wittgenstein has been obscured, Conant says, by its assimilation to Carnap's very different conception of meaning: 'Carnap seeks a method that will furnish criteria that permit one to establish that someone else is speaking nonsense, whereas Wittgenstein (both early and later) seeks a method that ultimately can only be practiced by someone on himself. Wittgenstein's method only permits the verdict that sense has not been spoken to be passed by the one who speaks' (ibid., p. 61). Carnap, of course, understood Wittgenstein as concerned also with objective meaning rather than with the mental states or intentions of speakers; Hacker (2003) thinks Carnap was right and Conant is wrong; see also Proops (2001).
3. Despite the Circle's (and especially Carnap's) intense preoccupation with this question in 1930–1, it soon emerged as an all-purpose, unanswerable one-line

*refutation* of logical empiricism. We still find it used this way, e.g. by Hilary Putnam: 'An obvious rejoinder [to the verification principle] was to say that the logical positivist criterion of significance was *self-refuting*: for the criterion itself is neither (a) "analytic"... nor (b) empirically testable. Strangely enough this criticism had very little impact on the logical positivists... I believe that the neglect of this particular philosophical gambit was a great mistake; that the gambit is not only correct, but contains a deep lesson' (Putnam 1981, p. 106).

4. 'Wittgenstein has sharpened the concerns about these three axioms by pointing out that they are not "tautologies"... Wittgenstein has given the impetus for further investigations by Russell himself and by Ramsey. However, his own conception differs from that of logicism ...' (Carnap 1930b, section 4).
5. It seems, indeed, that the two theories, Carnap's and Wittgenstein's, were inspired by very much the same models – Russell's logicist theory of relations on the one hand, and the structuralist view of physics (as expressed, for instance in the introduction to Hertz's *Mechanik*) on the other. The latter is specifically mentioned by Wittgenstein to motivate the idea of the 'multiplicity' (*Mannigfaltigkeit*) of a sentence in 4.032.
6. Mainly because of his 'construction principle', the quotation from Russell he used as an epigraph to the *Aufbau*: 'The supreme maxim of scientific philosophizing is this: Wherever possible, logical constructions are to be substituted for inferred entities.' For more details on this evolution see Carus (2007, pp. 161–77).
7. Kaila's critique (1930) was the first book-length assessment of the Vienna Circle; it focused its criticisms almost exclusively on Carnap's *Aufbau*. It frequently invokes Reichenbach (1929), which argues (pp. 26ff.) in favour of realism and against positivism, though not explicitly against Carnap. See Carus (2007, pp. 209–21).
8. This project is discussed in Awodey and Carus (2001), which also gives a more detailed account of the importance of the *Gabelbarkeitssatz* for Carnap's *Aufbau* project as well as the Vienna Circle's entire philosophy of logic and mathematics. Carnap's proof of the *Gabelbarkeitssatz* is actually correct, in his own terms, despite appearances. It does not, however, actually capture what he intended, as discussed in detail in the above paper, and as Carnap himself realized in 1930, even before Gödel's incompleteness results later that year.
9. In a manuscript from around this time, Carnap writes, 'If now, to introduce the infinite, one "adjoins ideal propositions" (Hilbert), i.e. writes down formulas that have no contentful [*inhaltliche*] meaning, but permit us to derive the mathematics of the infinite, then we have once again been able to determine the meaning of the signs introduced as meaningless, by investigating for which logical constants the formulas would become tautologies' (UCLA, RC1029/Box 4/CM13, item 2, p. 62).
10. In the margin of p. 3 of the manuscript, Carnap has scrawled, 'Regarding the undefined concept "*true*". It is completely different from the other concepts of metalogic. Perhaps avoidable? [Perhaps] just define which atomic sentences are the "basis" of a sentence, and how. (?)'
11. An addition of 7 February 1931 to the manuscript says, 'the syntax of the rows of dots is arithmetic' (p. 1).
12. The 'Attempt' ends with a summary in four points: '(1) The particular *natural numbers* occur as signs of *the language itself*. (2) The so-called "*properties of natural numbers*" are not proper properties, but syntactic (Wittgenstein: internal) ones, so

are to be expressed in the *metalanguage*. (3) A particular *real number* is a property or sequence of natural numbers, so is also to be expressed in the *metalanguage*. (4) The *properties of real numbers* are not real properties, but syntactic properties (with respect to the syntax of the metalanguage), and thus *to be expressed in the meta-metalanguage*' (p. 44).

13. In the Vienna Circle, he says, 'the philosophical problems in which we were interested ended up with problems of the logical analysis of language', and since 'in our view the issue in philosophical problems concerned the language, not the world', the Circle thought that 'these problems should be formulated not in the object language but in the metalanguage.' It was therefore 'the *chief motivation* for my development of the syntactical method' (our emphasis) to develop a 'suitable metalanguage' that would 'essentially contribute toward greater clarity in the formulation of philosophical problems and greater fruitfulness in their discussions' (Carnap 1963a, p. 55).
14. As Carnap explained in 'Die physikalische Sprache', not only criterial definitions but even ostensive definitions can be regarded as intra-linguistic. 'Elephant', for instance, criterially defined as an animal with certain characteristics, might be ostensively defined as 'an animal of the kind present at a certain space-time location' (Carnap 1932b, pp. 435–6).
15. The first systematic exposition of the new view was in a series of three lectures to the Vienna Circle in June and July of 1931. These fell into the period during which Rose Rand was taking minutes of the Circle meetings, so they are recorded, somewhat elliptically, in ASP, RC 081-07-17, 18, and 19 (with further discussion of these lectures in 081-07-20), and published in Stadler (1997, pp. 314–34). Carnap appears, from evidence in the file containing the 'Attempt', to have changed the system to an arithmetized one on 17 June, the day before the second lecture. He spells out the difference this makes to the scope of the system at the conclusion of the second lecture as follows: 'The difference between arithmetic metalogic and the metalogic portrayed previously is this: arithmetic metalogic treats not the empirically available, but all possible configurations. Our previous metalogic is the descriptive theory of certain given configurations, it is the geography of language forms, while the arithmetized metalogic is the geometry of language forms' (Stadler 1997, p. 325). Also noteworthy in these talks is the fact that they contain no definition of analyticity, and that they take the view that only a single language is required (something like the later Language I). In answer to the question 'So are we to draw the inference that there is only a *single* language?', Carnap replies 'Well, there are sentences of very different form ... but all of them, even the metalogical ones, are in a *single* language' (ibid., p. 329).
16. From the table of contents (ASP, RC 110-04-07) it seems clear that a single language (corresponding to the later Language I) was developed as the 'model Language' (*Modellsprache*). (In 'Die physikalische Sprache', it had been called the 'system language' (*Systemsprache*).) Just as in the June 1931 lectures to the Vienna Circle (see above, note 17) held just before Carnap embarked on composing the first draft, it seems that the 'model language' was regarded as the 'proper language' (*eigentliche Sprache*), while the full resources of classical mathematics could be developed by using the 'model language' as a metalanguage for axiomatic formal systems, Hilbert-style; the model language together with these axiomatic extensions was then called the 'extended model language'.
17. Presumably he refers here to Tarski (1932), which however gives only a bare summary; Gödel may have known more details from Tarski directly.

18. In a recent paper it is claimed that 'Carnap's main task in *Syntax* was to provide a reconstruction of mathematical truth' (Lavers 2004, p. 296; cf. also p. 308). This was true, as we have seen, of the first draft (the 'Metalogik') – but no longer of the published book. Note that the passages discussed by Lavers in support of his claim (ibid., pp. 297–9) were not included in the book's original edition (1934).
19. The classical exposition of this project is in Chapter 1 of *Logical Foundations of Probability* (Carnap 1950a); for further discussion, see Stein (1992), Awodey and Carus (2004), section III, and Carus (2004), section II.
20. There is widespread agreement about the continuity of Carnap's overall philosophical programme from the time of the *Syntax*; see e.g. Creath (1990), Ricketts (1996). Carus (2007) attempts a more systematic exposition of this later programme.
21. In which Gödel himself has, once again, played a posthumous role through the publication in 1995 of his critique of Carnap's view (Gödel 1995); see Ricketts (1994), Goldfarb (1995), Friedman (1999b), Potter (2000) as well as our discussions (2003, 2004) of Gödel's argument and these recent contributions.
22. Goldfarb and Ricketts (1992); Ricketts (1996, 2003); Friedman (1999b, ch. 9).
23. The second dogma is explicitly identified with the 'verification theory of meaning' in Quine (1951), p. 41.
24. Michael Friedman (forthcoming) discusses Quine's assimilation of Carnap to a Humean empiricism, and his persistent reading of both *Aufbau* and *Syntax* in this light; even as early as Quine's 1934 lectures on logical syntax at Harvard (Creath 1990, pp. 47–103), the principle of tolerance is not mentioned. It seems likely that Quine went on attributing to the published *Syntax* the (pre-tolerance) view reflected in the first draft, which he had read (Quine 1986, p. 12) as it was being written, in Prague. This would explain his failure to understand the centrality of tolerance to Carnap; it does not explain, however, how Quine could have missed Carnap's evident rejection of the *second* dogma in the first draft of the *Syntax*. Perhaps it resulted from the slowness of the process by which Carnap adjusted his epistemological views to the new syntax doctrine; see Carus (2007, pp. 39–45).
25. Friedman (forthcoming) cites examples, including the introductory paragraph to the original publication of 'Carnap and Logical Truth', omitted on republication in *The Ways of Paradox*.
26. According to Howard Stein's recollection of the discussion after a colloquium talk Quine gave in Chicago in 1951, Carnap summarized the differences between himself and Quine (a summary with which Quine, at the time, concurred) approximately as follows: 'Quine... and I really differ, not concerning any matter of fact, nor any question with cognitive content, but rather in our respective estimates of the most fruitful course for science to follow...' (Stein 1992, p. 279).

# Part II

# Philosophy of Mathematics and Logic

# 3

# Carnap's *Syntax* Programme and the Philosophy of Mathematics

*Warren Goldfarb*

In the period leading up to *The Logical Syntax of Language*, from about 1928 onwards, logic and the foundations of mathematics were a principal focus of Carnap's thinking. On the more technical side, he published the *Abriss der Logistik* in 1929, and started the ill-fated project called *Untersuchungen zur allgemeinen Axiomatik* in 1928.[1] 1930 saw the publication of a report on this project (1930d), and, on the more discursive side, of 'Die alte und die neue Logik' (1930a) and 'Die Mathematik als Zweig der Logik' (1930b), as well as the delivery of his address 'Die logizistische Grundlegung der Mathematik' (1931a) in September at Königsberg. Those years also saw conversations between Carnap and Gödel, including discussion of the incompleteness theorem as early as August 1930, and between Carnap and Tarski. All this activity is the background of the 'sleepless night' in January 1931 during which a fevered Carnap conceived of the central idea of *Logical Syntax*, as he reported in his 'Intellectual Autobiography' (1963a, p. 55).

The heated controversies in foundations of mathematics of the 1920s were central to Carnap's motivations in writing *Logical Syntax*. When he addresses those controversies in the Foreword to the book, he talks of 'pseudoproblems' and 'wearisome controversies'; to eliminate these 'is one of the chief tasks of this book' (1934d/1937, pp. xiv–xv). Indeed, he claims the procedures recommended by *Syntax* have the effect that 'the dogmatic attitude which renders so many discussions [in the foundations of mathematics] unfruitful disappears' (§16). Thus, the position on the foundations of mathematics presented in *Syntax*, Carnap suggests, will end the wrangling of the foundational schools.

In this chapter I shall discuss how Carnap's adoption of the stance of *Logical Syntax* alters his own earlier views on logic and foundations of mathematics: what he retains, what he willingly changes, and what he winds up having to change further, willy-nilly perhaps, pushed by the technical situation he had himself set up. I start by describing Carnap's position in the period immediately preceding *Syntax*. At that time many of the technical issues were not all that well understood. Carnap's views reflect this, but (perhaps for that very reason) they also display his philosophical inclinations more clearly. His stance toward then-current positions is made explicit in the two papers 1930b and 1931a. The latter is fairly well-known, as it appeared in English translation in the widely used anthology by Benacerraf and Putnam (1964). The former is much less familiar: it was published in the nationalistic journal *Blätter für deutsche Philosophie* (in fraktur!) and never reappeared, neither anthologized nor translated. Like 1931a, it is a contribution to a symposium on foundations of mathematics, with the different schools represented; but it is framed in more elementary terms than the better-known paper, and so gives a more direct statement of where Carnap sits.

In Carnap's writings of these years, he adopts Wittgenstein's notion of tautology, and stresses the truth-table analysis of logical truth, the idea that the propositions of logic are tautologies in that they rule out no possibilities. Unlike Wittgenstein, though, Carnap wants to extend this characterization to mathematics, by means of the logicist reduction. He urges in favour of logicism: 'Mathematics is a branch of logic [...] The concepts of mathematics can be derived from logical concepts [...] The propositions of mathematics form part of the propositions of logic.' Yet he presents himself as 'an investigator who acknowledges the basic idea of logicism, but who is critical of the extant attempts at carrying it out' (1930b, pp. 298 and 308).

Carnap's criticism of the current logicist reductions have entirely to do with the use of impredicative definitions. Now, Carnap sought to avoid ramification in the theory of types, for several reasons. First, he agreed with Ramsey in distinguishing the semantical paradoxes from the logical paradoxes, in thinking that the logical framework need address only the latter, and in seeing that the simple theory of types does so adequately. Second, he agreed with Ramsey and Wittgenstein that the axiom of reducibility is not a logical principle; obviously, then, logicism would not have a chance at success unless ramification is discarded. Third, he found the basic structure of the simple theory of types could be motivated entirely apart from the need to avoid paradoxes:

> The simple theory of types is to be accepted. The principal reason for this is less the avoidance of the antinomies – which of course were the original impetus to the formulation of the theory – than the circumstance that in the construction of a system of concepts of empirical science the division into types and the nonsensicality of crossed-types come about automatically, so to speak. (1930b, p. 308)

Here he is clearly reflecting on some features of his work in the *Aufbau*.

The naturalness of the simple theory of types was urged by other authors in those years. In both Fraenkel (1928) and Hilbert and Ackermann (1928), the type distinctions of the simple theory are clearly marked out as basic, and Whitehead and Russell's system is characterized as having 'further constraints'. Thus both texts, unlike *Principia Mathematica* itself, treat the hierarchy of orders as superimposed on a hierarchy of types; this quickly becomes the standard view. It also leads to the widespread discarding of ramification. By late 1930, it appears, the terminology 'simple theory of types' had become entrenched; and in 1931a Carnap could claim that 'most proponents of modern logic consider [the simple theory] legitimate and necessary' (p. 97).

For Carnap the problem then arises of how to justify the impredicativity that the simple theory allows, that is, how to show that the impredicative comprehension principle is tautologous. He is emphatic in rejecting Ramsey's view, that 'mathematics has to do not only with those classes that are defined by specifiable [*angebbaren*] concepts'; Ramsey's conception of 'non-definable [*nicht-definierbaren*] logical structures' is not to be accepted (1930b, pp. 307–8). Carnap thus expresses a restrictive view of the relation between legitimate mathematical entities and specifiability (probably more restrictive than Russell's own). At this point, too, Carnap uses the epithet 'theological mathematics' to characterize Ramsey's proposal, a rubric he repeats in 1931a. It is not, in Carnap's mouth, a commendation.

Indeed, the justification of impredicative definition is for Carnap the *only* problem that remains for logicism. With respect to the axiom of infinity and the axiom of choice, he simply adopts Russell's expedient of adjoining the axioms as additional antecedents of a conditional, so that, for example, Euclid's theorem on the infinitude of the primes becomes translated into logistic as 'If there are infinitely many individuals, then there are infinitely many primes'.

Finally, in this paper Carnap turns to a brief assessment of the other schools in foundations of mathematics. Of course he rejects the intuitionists' invocation of intuition. But he commends their proposed limitation to only those existence proofs that involve construction procedures, and concludes:

> The logicist will also have to recognize the requirement of constructivity, and as a result must undertake a critical reexamination of the system of *Principia Mathematica*: that means proceeding in a direction directly opposed to Ramsey's absolutism. (1930b, p. 309)

The suggestion, I take it, is that there should be a constructive method that would supplant Ramsey's method of justifying impredicative definition of sets. Carnap does not give any indication of what this might be.

With respect to Hilbert-style formalism, Carnap finds a gap, in that treating mathematics as signs without meaning does not do justice to the applications of mathematics: to show that applications in the empirical world are logically legitimate, interpretations have to be given to the mathematical signs. Thus formalism has to be supplemented; and, so supplemented, there are grounds for thinking that the meanings of the number-signs will agree with those given them by logicism. As Carnap put it in the discussion after his lecture in Königsberg, 'The formalist introduction of the natural numbers would receive a logicist interpretation' (Carnap 1931c, p. 143). Carnap concludes:

> The details of a definitive solution of the problem of the foundations of mathematics cannot yet be anticipated [...] It may well be that the discovery of a solution to the problem which will appear satisfactory from all the different viewpoints can no longer be viewed as so much without prospects as it appeared just a few years ago. (1930b, p. 310)

Thus, Carnap ends with an expression for the hope of consilience of logicism, formalism, and intuitionism. This is a characteristically Carnapian move, to try to reconcile philosophical positions that seem opposed. For example, in *Der Raum* (1922), Carnap wants to show that apparently incompatible viewpoints about space are in fact compatible, because they are talking about different spaces; a proper distinction of those spaces allows all the positions to be accommodated. In the *Aufbau* (1928a, §178), Carnap wants to show that realism, idealism, and phenomenalism can be put into harmony, once they are purged of their metaphysical components.

Here, however, Carnap does not give any real grounds for his optimism. He does note that formalism has adopted a stringent requirement of finitism in carrying out metamathematics, and thus 'has brought into its system a recognition of the most important basic idea of intuitionism'. (He does not remark that Brouwer rejected Hilbert's programme completely.) Carnap is mute on how it might happen that intuitionists and logicists could find a place for rapprochement, except for the hope for a 'critical reexamination' of *Principia* that would, at one and the same time, be constructive and yield the acceptability of impredicative definitions.

In 1931a, Carnap is not as unguarded, but the basic viewpoint seems to be the same. The only difficulty that Carnap sees in the logicist position on foundations of mathematics is the question of impredicative definition. He repeats his characterization of Ramsey as 'theological mathematics' and adds 'Such a conception, I believe, is not far removed from a belief in a Platonic realm of ideas which exist in themselves' (p. 102). But in this paper Carnap actually gives a substitute for Ramsey's 'theological' considerations, or at least the suggestion of one: an attempt at giving an affirmative answer to his questions 'can we have Ramsey's result without retaining his absolutist conceptions?' and 'can we allow impredicative definition [...] without falling into [Ramsey's] conceptual absolutism?' (p. 103).

What follows then is an attempt at some kind of legitimation of impredicative definition. It is essentially an argument that, in a specific case (the definition of the finite numbers), impredicativity does not lead to a vicious circle. The question of circularity is taken in an epistemological way, unlike Russell's more ontological concerns when he framed the vicious circle principle in 1905.[2] Carnap asks, can we in a non-circular way *derive* results about the applicability of impredicatively defined properties? His example is the property of being an inductive number, that is, that of being a number that possesses every hereditary property possessed by 0. The problem is that among the hereditary properties is that of being an inductive number itself. Nonetheless, Carnap shows, it is possible to show with circularity that various numbers are indeed inductive. There is much to be said about this argument, particularly Carnap's interpretation of the threat of a vicious circle and what needs to be done to show it can be averted. Moreover, Carnap gives only this one example. His argument is that in this one case the impredicative definition poses no danger. He makes no attempt to handle impredicative definition generally. (This one case has an especially simple logical structure. To be sure, it is the most central case for the logicist reduction of arithmetic. But it is not at all clear how Carnap's argument would extend to the more complicated impredicative definitions that would figure in formalizing the theory of real numbers within the theory of types.)

For my purpose here, though, the important thing to take from Carnap's discussion is the simple fact that, at this stage, Carnap thinks he needs some kind of argument to legitimize impredicative definition. It isn't exactly a justification, but it is at least a way of showing that such definitions do not (or do not necessarily) lead to trouble. That is, Carnap feels compelled to have some substitute for Ramsey's invocation of arbitrary infinitely complex propositional functions that are independent of our specifications. Moreover, clearly Carnap thinks more work is needed. He expresses some tentativeness: immediately after this argument, he says, 'If this theory is in fact feasible [...]' (p. 104).

Carnap ends 1931a with another conciliatory gesture. 'Logicism as here described has several features in common both with intuitionism and with formalism' (p. 104). He does not repeat the hope for a solution that all schools will accept, but rather simply characterizes what he finds as commonalities between them.

The characterization of the links between logicism and intutionism is particularly revealing. Logicism, he says,

> shares with intuitionism a constructivistic tendency with respect to definition [...] A concept may not be introduced axiomatically but must be constructed from [...] primitive concepts step by step through explicit definitions. (p. 105)

He credits Frege with urging this. The position is one he had written on more at length in 'Eigentliche und uneigentliche Begriffe' (1927), a paper on explicit and implicit definitions that is clearly directly influenced by Frege's disapprobatory views of the latter. Vis-à-vis intuitionism, however, what is noteworthy is the distinction that Carnap is *not* making. Earlier in 1931a, he had praised Dedekind's construction of the real numbers. It appears, then, that Carnap understands by 'construction' something far less restrictive than what the intuitionists were urging. This is confirmed in the next sentence:

> The admission of impredicative definitions seems at first glance to run counter to this tendency, but this is only true for constructions of the form proposed by Ramsey. [...] The difference between us [and the intuitionists] lies in the fact that we recognize as valid not only the rules of construction which the intuitionists use (the rules of the so-called 'narrower functional calculus'), but in addition permit the use of the expression 'for all properties' (the operations of the so-called 'extended functional calculus'). (p. 105)

Now, the narrower functional calculus is Hilbert and Ackermann's name for first-order quantification theory – *classical* first-order logic, that is. So Carnap is mischaracterizing the intuitionists as allowing all classical first-order constructions. (For example, this would allow the construction of a path through a recursively specified infinite finitely-branching binary tree, even when this is not possible constructively.) Indeed, his characterization best fits not Brouwer's intuitionism but what are sometimes called the 'quasi-intuitionists', usually meaning Poincaré, Borel, and the Weyl of *Das Kontinuum* (that is, Weyl *before* he joins forces with Brouwer), since they all accepted classical logic, but rejected impredicative definitions.

It does not appear, then, that at this juncture Carnap has much understanding of intuitionism proper, despite having heard Brouwer lecture in 1928. This is not difficult to understand. Brouwer's papers were intricate and obscure. The technical extent of intuitionist principles was not clarified until later in the 1930s, through the work of Heyting and Gödel. Before the mid-1930s, no distinction was generally made among 'finitary', 'constructive', and 'intuitionist'. Carnap may also have been influenced in his characterizations by Felix Kaufmann's *Das Unendliche in der Mathematik und seine Ausschaltung* (1930), which he reviewed (1930c) and cited in both 1931a and *Logical Syntax*. Kaufmann, it seems, had his own brand of constructivism, rejecting the uncountable, but allowing such things as the theory of countable well-orderings. (In 1930c Carnap writes 'These basic laws suffice for set-theoretic topology, as well as for classical analysis.' So clearly the position includes far more than standard intuitionism.)[3]

To sum up: in the period just before *Syntax* Carnap espoused a rather traditional logicism, based on the logicist reduction of real numbers to Dedekind cuts and of the natural numbers to classes of equinumerous classes delimited by the ancestral, where the framework logic is the simple theory of types. He waved off the difficulties of the axioms of infinity and choice, by conditionalizing them, but did feel he needed to provide something to underwrite the impredicative definitions that the simple theory allows, because he still thought of logic as dealing solely with specifiable entities, in some sense of specifiable.

The move to *Syntax* freed Carnap up from the constraints this position put on him. The basic contrast can be illustrated thus: at the end of the 1931a he says 'Logicism proposed to construct the logical-mathematical system in such a way that [...] the axioms and rules of inference are chosen with an interpretation of the primitive symbols in mind'; while in the Foreword to *Logical Syntax* he says:

> Up to now, in constructing a language, the procedure has usually been, first to assign a meaning to the fundamental [...] symbols, and then to consider what sentences and inferences are seen to be logically correct in accordance with this meaning [...] no conclusion arrived at in this way can very well be otherwise than inexact and ambiguous.

And he continues that the matter has to be approached from the opposite direction:

> Let any postulates and any rules of inference be chosen arbitrarily; then this choice, whatever it may be, will determine what meaning is to be assigned to the fundamental [...] symbols. (1934d/1937, p. xv)

This is a direct reversal of the Fregean position explicitly argued in Carnap (1927), and signalled in Carnap (1930b) and (1931a) by Carnap's insistence on constructions rather than axiom systems.

Underwriting this shift is Carnap's pluralism about languages: there are alternative linguistic frameworks, many different logics of inference and inquiry. Since justification can proceed only grounded in the logical relations of a particular framework, justification is a notion internal to each specific language. Thus there can be no question of justifying the choice of one language over another. Carnap frames his pluralism in his principle of tolerance:

> *In logic there are no morals*. Everyone is at liberty to build up his own logic, i.e., his own form of language, as he wishes. All that is required of him is that [...] he must state his methods clearly, and give syntactical rules instead of philosophical arguments. (§17)

What emerges from this sea-change is Carnap's true original contribution to philosophy of mathematics: a version of logicism that does not require the logicist reduction. Residual worries about the axiom of infinity can fall away, since the natural numbers can be introduced on their own (and presumably the real numbers can be treated similarly). The general idea of 'language' takes the place of any specific system of logic. Consequently, Carnap's logicism becomes merely a commitment to the elaboration of a linguistic framework containing both pure mathematics and the means to apply mathematics to the empirical world. The question of whether mathematical concepts are to be defined in a vocabulary Carnap calls 'logical in the narrower sense' – that is, the vocabulary of *Principia Mathematica* – rather than taken as primitive 'is not a question of philosophical significance, but only one of technical expedience' (§84).

The problem of impredicativity, which had loomed so large in 1930b and 1931a, has essentially disappeared. Carnap discusses the issue in §44:

> The proper way of framing the question is not 'Are impredicative symbols admissible?', for since there are no morals in logic (see §17), what meaning can 'admissible' have here? The problem can only be expressed in this way: 'How shall we construct a particular language? Shall we admit symbols of this kind or not? And what are the consequences of either procedure?'

Carnap no longer feels any need to have a substitute for Ramsey's kind of justification. His language is absolutely forthright here:

> The material reasons so far brought forward for the rejection of [...] impredicative terms are not sound. We are at liberty to admit or reject such definitions without giving any reason. But if we wish to justify either procedure, we must first exhibit its formal consequences.

This is not to say that the argument from 1931a is completely without point. Carnap does cite it; but its role has changed. It is no longer an essential justification; it is invoked just to show that languages which allow impredicative definitions may also contain means for determining whether the impredicative term is applicable in an individual case. The difference is small, but telling. For example, in the *Syntax* context, I think, there is no necessity for trying to extend this argument to impredicative definitions of more complexity. Moreover, Carnap thinks the argument is not all that important, since the requirement that we should use only languages in which there is the possibility of determining whether the impredicative term applies in an individual case is 'too narrow, and [...] not convincingly established'.

Indeed, the *Syntax* viewpoint, with the principle of tolerance at its heart, involves a real shift in attitude toward the constructivist criticisms. Most importantly, consilience is dropped: it simply disappears as a goal. There is no longer any motivation to try to accommodate the different stances in foundations of mathematics in one view. That is, tolerance is used *not* as a furtherance of the conciliation approach that is pervasive in earlier

Carnap. Rather, tolerance makes conciliation unnecessary. The foundational wars are to be ended not by finding one solution that satisfies everyone, but by allowing everyone to exhibit the virtues (or disadvantages) of his own approach. But the exhibition is limited to the precise technical features of the view, that is, the actual mathematico-logical principles that are being urged.

Indeed I find in Carnap's prose at this point a sense of impatience with intuitionism: that much of what the intuitionists say is just gas, whereas the important thing is to display the formal characteristics of the proposed system. This, combined with Carnap's slogan 'It is not our business to set up prohibitions.' (§17) amounts to an anti-Brouwerian tone quite distinct from the irenic tendencies of the previous years. By this time, Carnap also has a better understanding of the distinctions among types of constructivist views. For example, he marks out the distinction between those who do not want to use undecidable predicates (the finitists), and those who bridle only at impredicativity. I suspect that Carnap's coming to understand the multiplicity of such views, and how they differ one from another, is another motivating factor for the whole *Syntax* approach. In the end, what the procedure of *Logical Syntax* enables Carnap to do (or to have the conceit of doing) is to transform philosophical questions into technical, mathematical questions; but unlike Hilbert's attempt at similar transformation, to allow all schools to participate.

This open invitation, however, should not be confused with neutrality. Even the insistence on precise axiomatics tips the balance against Brouwer. Most importantly, the tolerance-based approach refuses to recognize a central strain in the intuitionistic criticism of classical reasoning, in which it is alleged that the classical understanding of the connectives is, in some way, incoherent; that is, for example, there is no consistent understanding of 'or' and 'not' that supports the law of excluded middle. Carnap explicitly rules out any idea that logical laws need to be supported. Hence this line of criticism is, from the *Syntax* point of view, a non-starter.

Carnap did claim to see some virtue in intuitionism. He expresses this by formulating Language I in *Logical Syntax*, which is a constructive language, with all quantifiers bounded. This represents the most tractable part of a constructivity requirement, since it can formalize only finitary reasoning. Carnap never acknowledges that there are further realms of *constructive* reasoning, and even more wide-ranging realms of intuitionistic reasoning (I have in mind here such results as the Fan Theorem, which are peculiar to intuitionism). Clearly he'd prefer it if we do whatever mathematics we can as finitistically as we can. (Who wouldn't?) But there is no further commitment to the virtues of constructivity. It is also interesting to note that Carnap treats Language I metalinguistically in a classical way (so that, for example, free-variable statements, which express unbounded generality in this language, are taken, in a certain sense, to be bivalent).

The arena Carnap sets up in *Logical Syntax* for discussion of the foundations of mathematics can look like paradise: within it there appear to be no-cost solutions to the problems that beset all of the approaches of the 1920s. There is, however, a serpent in the garden. Gödelian incompleteness forces the formulation of rules of consequence, over and above rules of deduction, in order to obtain classical bivalence. Carnap finds this easy to treat for the first-order language. But for the higher-order language that allows impredicative definitions, Carnap's attempt at formulating those rules of consequence runs into an obstacle that Gödel pointed out to him. This is not, or not just, a question of the strength of the metalanguage that is needed. It concerns more directly the very possibility of using only truly language-based entities in the definition of logico-mathematical truth.

To explain, let me proceed historically, and track through Gödel's criticism and Carnap's reaction.[4] In 1932 Carnap sent Gödel a draft called *Semantics*, which was the word Carnap was then using for what he later calls 'syntax'. In it, Carnap gave rules of consequence for impredicative higher order logic by attempting to define what it was for sentences of this language to be analytic. In modern terminology, Carnap was attempting to give a truth definition for the language. In his definition, Carnap had framed the condition for analyticity of a higher-order universal quantification in terms of the analyticity of all instances of the quantification, where instances are obtained by replacing the higher-order variable with predicates. In a letter of 11 September 1932, Gödel gave a lethal criticism of Carnap's definition: since the predicates can be impredicative, an instance of a quantification can contain just the same quantifier within it, and so the definition is viciously circular. Gödel wrote:

> In my judgment, this error may only be avoided by regarding the domain of the function variables not as the predicates of a definite language, but rather as all sets and relations whatever. On the basis of this idea, in the second part of my work I will give a definition for 'truth', and I am of the opinion that the matter may not be done otherwise, and one can *not* view the higher functional calculus semantically. That is, one can of course build up a higher functional calculus on a semantic basis, but then just those laws that one needs for the classical theory of the real numbers are not satisfied, because one is led of necessity to ramified type theory (without the axiom of reducibility).

Carnap responded, on 25 September 1932:

> I realize that one will not arrive at the formulas of classical mathematics about real numbers if the universal quantifier with predicate variables (or function variables) ranges only over the predicates definable in a definite delimited system. You say: it must range over 'all sets'; but what does that mean? [...]

> If the definition is to achieve what was striven for, then for the rule [...] one may take not a restricted semantic language; rather the rule may be constructed with arbitrary semantic notions. (For otherwise certain sets of numbers would again remain outside, sets that are indeed plausible, but that cannot be comprehended within the system.) But is this not objectionable?

Two days later, Carnap wrote that he found 'the solution' to his difficulty:

> The locution 'for every valuation' that occurs in the definition can still be expressed in a semantics formulated in a definite language, namely by '(F)(...)', since a valuation is of course a semantic predicate. This is possible even though in the semantics under consideration, not all possible valuations, that is, predicates, can be defined.[5]

Thus we see Carnap adopting a position which goes directly against the theme of the definability of all entities that he had sounded in both 1930b and 1931a. Note how Carnap is forced to this. One could think that even impredicative definitions are still compatible with the notion that one is dealing only with specifiable entities: they would be specifiable in the extended functional calculus, that is, in higher-order logic. But in order to carry out the *Syntax* programme, Carnap needs to give rules of consequence – which amount to a semantics in our sense – that will yield impredicative definitions, and, as he sees from Gödel's objections, that will require giving up the requirement of specifiability.

The upshot of this is visible in *Logical Syntax*, §34d:

> Thus the definition must not be limited to the syntactical properties which are definable in S, but must refer to all syntactical properties whatsoever. But do we not by this means arrive at a Platonic absolutism of ideas, that is, at the conception that the totality of all properties, which is non-denumerable and therefore can never by exhausted by definitions, is something which subsists in itself, independent of all construction and definition? From our point of view, this metaphysical conception – as it is maintained by Ramsey for instance [...] – is definitely excluded. We have here absolutely nothing to do with the metaphysical question as to whether properties exist in themselves or whether they are created by definition. The question must rather be put as follows: can the phrase 'for all properties...' (interpreted as 'for all properties whatsoever' and not 'for all properties which are definable in S') be formulated in the symbolic syntax-language S? This question may be answered in the affirmative. The formulation is effected by the help of a universal operator with a variable... i.e., by means of '(F)(...)', for example.

So here Carnap is simply denying that one needs anything like Ramsey's arguments about ontology, or any substitute for it.

The question is: how can Carnap help himself to this, flying, as it does, in the face of his previously held position? The answer is, clearly, the principle of tolerance. So I agree with Awodey and Carus that it was Carnap's interchange with Gödel in 1932 that moved Carnap to formulate the principle of tolerance. (See their contribution to this volume.) But I differ from them as to the reason. It is not, to my mind, an issue of the plurality of metalanguages that motivates Carnap. Rather, it is the need to have an opening for the view that logical syntax can use notions not specifiable in a particular system without thereby committing itself to Platonism, infinitarism, or the like. 'In logic, there are no morals.' gives Carnap that opening.

At the same time, Carnap does not face up to the consequences. If the use of an unbridled universality operator over higher order objects does not bespeak a Platonistic commitment, it must be connected somehow to convention. But how does this convention get to be determined? There seems to be no way to do this, except to say that it's a matter of the meta-metalanguage. The nature of that language is then settled only given the nature of the meta-meta-metalanguage. And so on. This is not an incoherent position; it is, as I have written elsewhere, 'self-supporting at each level'. But it does have more than a whiff of circularity or at least of vacuity, which, of course, Carnap's critics will exploit.

## Notes

1. The nature of this project, and Carnap's reasons for abandoning it, are discussed in Awodey and Carus (2001). See also Goldfarb (2005).
2. See Goldfarb (1989).
3. Mancosu (2002) notes that in this book Kaufmann conjectures that every classical proof of an existence theorem, provided it relies on no non-constructive existence axioms like the axiom of choice, contains in some implicit form a construction of an instance; and that Kaufmann later claims a proof of this conjecture. Gödel then formulates a counterexample that shows Kaufmann to be completely wrong. Kaufmann's conjecture may indicate that consilience between classical and constructivist views was generally in the air in Vienna.
4. These letters are in Gödel (2003, pp. 347–57).
5. In his reply, dated 28 November, Gödel wrote that Carnap had correctly understood what he meant about the definition of 'analytic'. He continues: 'I believe moreover that the interest of this definition does not lie in a clarification of the concept "analytic", since one employs in it the concepts "arbitrary sets", etc., which are just as problematic.' This may be the only time that Gödel expresses a position that is less ontologically exuberant than Carnap's.

# 4

# Carnap on Logical Consequence for Languages I and II

*Philippe de Rouilhan*

## 1 Prologue: Carnap in search of a SYNTACTICAL explication of the relation of logical consequence

In his 'Intellectual Autobiography', Carnap recalled what Bergmann would call the 'linguistic turn' (1964, p. 177) in philosophy in the days of the Vienna Circle:

> In our discussions in the Vienna Circle it had turned out that any attempt at formulating more precisely the philosophical problems in which we were interested ended up with problems of the logical analysis of language. (1963a, p. 55)

That is how philosophy became the logical analysis of language. More precisely, for Carnap, at the time of *Syntax*, philosophy was SYNTACTICAL, and only SYNTACTICAL, analysis of the LANGUAGE of science. (Neither 'semantic' nor 'pragmatic' analyses were yet on the agenda.) At that time, in contrast to ours, SYNTAX as such did not imply any restriction upon the mathematical resources invested in the undertaking, and a LANGUAGE as such not only involved rules of *formation*, but also rules of *transformation*.[1]

The most fundamental concept of SYNTAX is the relation of logical consequence, that it is a question of explicating. Connected with this relation are the concepts of logical validity (or analyticity), of logical contravalidity (or contradiction), of logical determination (or L-determination, viz., analyticity or contradiction), of logical indetermination (or syntheticity), of logical content, etc. These concepts are mobilized in the formulation of philosophical propositions – or rather proposals – such as: sentences of mathematics are logical; those which are valid are logically valid, analytic, void of logical content; the sentences of empirical sciences are synthetic; and there is nothing in between the two; etc. What is philosophically at stake in an adequate explication of the concept of logical consequence is therefore considerable.

The aim of the present chapter is to shed some light on the astonishing explications of the concept of logical consequence that Carnap proposed for LANGUAGES I and II given as examples in parts 1 and 3 of *Syntax*. I shall not call into question Carnap's point of departure, namely the distinction presupposed between logical and extra-logical (*'descriptive'*) symbols, nor shall I say anything about the inverse attempt (developed in part 4 of the book) to base that distinction in a completely general way on the concept – assumed to be given prior to any distinction between logical and extra-logical – of direct consequence. Carnap must be given credit for having been the first to seek an explication of the concept of logical consequence as being different from just any concept of derivability. It is probably in reaction to this attempt that Tarski published his 1936 article 'On the Concept of Logical Consequence' (Tarski 1936b). The principal constraint weighing upon Carnap's explications and making them so troubling for the readers of Tarski that we are is that the concept of truth is banned from them, while it is in a certain way at the heart of the most commonly accepted informal explanation.[2] Where Carnap saw himself forced into useless and costly contortions, Tarski, armed with his famous definition of the concept of truth, would only need a few crystal-clear pages to succeed in the same undertaking.

## 2 Between Gödel's theorem and Tarski's *Wahrheitsbegriff*

### 2.1 Before Tarski's *Wahrheitsbegriff*, the standard, informal explanation in terms of truth seemed not to be convertible into a SYNTACTICAL explication

The informal explanation of the relation of logical consequence that I shall call standard seems clear: *a class of sentences (the premises) has as* ***logical consequence*** *a sentence (the conclusion) if, and only if, for any interpretation of the extra-logical symbols involved, if the premises are true (for this interpretation), then the same holds for the conclusion*. There is nothing at all unusual about that informal explanation of logical consequence, and I am prepared to make it go back to Aristotle, when he defined syllogism at the beginning of *Topics* and schematized it at the beginning of *First Analytics*. On the informal level, this is, I believe, how Carnap understood it. See, for example, the last lines of §12 of *Syntax*:

> [The rules of inference of Language I] are formulated in such a way that, when materially interpreted, they always lead from true sentences to further true sentences. (LSL, p. 32)

Admittedly, not all sentences of LANGUAGE I are closed, but, except when they are in the position of subsentences, open sentences of that LANGUAGE must be construed as versions of their universal closures and thus treated as if they were closed. The same holds for LANGUAGE II. So it is that the standard, informal explanation holds for all sentences of Carnap's LANGUAGES I and II, be they open or closed.

Let me remove possible misgivings about my reading of the quoted sentence. Did not Carnap have in mind variation of extra-logical symbols (according to their syntactic category) rather than that of their interpretations? Indeed, at the informal level, both variations are not always clearly distinguished. However, they intuitively amount to the same thing if, as Carnap did, one assumes the possibility for the extra-logical vocabulary of the LANGUAGE under consideration to be enriched *ad libitum*. For, then, it intuitively comes to the same thing to take into account all replacements of extra-logical symbols by whatever *possible* symbols (according to their syntactic categories) or all replacements of their interpretations by whatever interpretations.

In his *Wissenschaftslehre* (Bolzano 1837), Bolzano had proposed a comparable explication, but it was ontological, not SYNTACTICAL, in character, for it concerned propositions-in-themselves (*Sätze an sich*), not sentences. In 1936, Tarski would return to the standard, informal explanation, secured *in a certain way* (see below) by the explication of the concept of truth that he had given earlier in the *Wahrheitsbegriff* (1933a/1935). But, in 1934, Carnap did not yet know Tarski's work on truth (the 1933 version was in Polish) and did not see how it would be possible to explicate this concept in SYNTACTICAL terms. He could not, therefore, resort to the standard, informal explanation and had to look for an alternative.

Let us take another, more finely-tuned, clearer look at this story. A distinction definitely has to be drawn between, *on the one hand*, the monadic concept of *absolute* truth, applying to sentences of an interpreted object-LANGUAGE (the interpretation of which being possibly given through its translation into the meta-LANGUAGE, as for Tarski in the *Wahrheitsbegriff*), and, *on the other hand*, the dyadic concept of *relative* truth, applying to sentences of a non-interpreted object-LANGUAGE and applying to them relative to some interpretation or other of this object-LANGUAGE, or, if one prefers, applying to sentences of an interpreted object-LANGUAGE and applying to them relative to some reinterpretation or other (possibly the identical one) of this object-LANGUAGE. When speaking of an object-LANGUAGE being interpreted, non-interpreted, or reinterpreted, I am exclusively referring to extra-logical symbols. As for logical symbols, their 'interpretation', so to speak, is supposed to be SYNTACTICALLY fixed, once for all.

Carnap would be quite right to doubt that an explication of truth was possible in SYNTACTICAL terms, if it were a question of absolute truth. Tarski's work could not do anything about that, *pace* Tarski himself and Church, who would unwisely affirm the opposite.[3] In the *Wahrheitsbegriff*, Tarski by no means gave a SYNTACTICAL definition of the concept of absolute truth, because, in general, that is just simply impossible. But, the standard, informal explanation of the concept of logical consequence by no means mobilizes the absolute concept of truth, only that of relative truth (in the formulation of the standard, informal explanation given above, this relativity is made explicit by recourse to parentheses). Now, from Tarski's original work on the concept of absolute truth, it is not very difficult to extract a purely SYNTACTICAL explication of the concept of relative truth and, finally, as Tarski would do in 1936 (admittedly, without worrying about defining the concept of relative truth, nor, a fortiori, about defining it SYNTACTICALLY), of the concept of logical consequence. Carnap was, therefore, wrong to doubt that a SYNTACTICAL explication of the concept of truth was possible, if it was just a question of the concept of truth – the concept of relative truth – needed by the standard, informal explanation of the concept of logical consequence, and he was wrong finally to consider this last explanation to be irreducible to a SYNTACTICAL explication.

On the whole, what Carnap lacked, it seems, was first of all a recognition of the distinction between the two concepts of truth and the correct identification of the one involved in the standard, informal explanation of the concept of logical consequence. Or, to put it less bluntly, having correctly identified the concept of truth involved as being that of relative truth, but not having sufficiently distinguished it from the concept of absolute truth, Carnap no doubt projected the same highly problematic, paradoxical, and SYNTACTICALLY irreducible quality onto the first that he had perceived in the second (even though he did not go so far as to ban the latter as metaphysical and meaningless, as certain of his friends in the Vienna Circle were doing), and consequently believed that he had to abandon recourse to the standard, informal explanation.

## 2.2 Gödel's work prohibits falling back upon a SYNTACTICAL explication in terms of derivability

If the relation of logical consequence, or 'L-consequence', must be explicated, or defined, for a LANGUAGE whose rules of transformation are all taken to be *logical*[4] in the sense of governing the use of logical symbols in conformity with their intended meaning, and are *finitary*,[5] like, for example, Carnap's rules of *derivation* of LANGUAGES I and II, and, if this is to be done without resorting to the concept of truth (be it but relative), one cannot help but think of doing this in terms of these rules, that is, in terms of derivability. Since Gödel (1931), we know that, under certain very general conditions, such an explication will necessarily be incomplete, and when writing *Syntax* Carnap knew this well. But, let us imagine Carnap writing *Syntax* before Gödel. He has at his disposal concepts of derivability and related concepts of demonstrability, refutability, decidability (*'resolubility'*), etc., and he uses them to explicate the relation of L-*consequence* and related concepts of L-*validity* (or *analyticity*), L-*contravalidity* (or *contradiction*), L-*determinacy*, L-*indeterminacy* (or *syntheticity*), etc.

The question then arises of the adequacy of such an explication, that is, of its soundness and its completeness. However, the adequacy between an informal *explicandum* and its formal *explicatum* does not itself admit of formal proof. As for soundness, one can at least have an intuitive sense of it by examining the rules of derivation one by one. As for completeness, one must be content, for want of something better, to assume it, until one finds proof to the contrary. Direct proof of the contrary would consist in finding a rule of transformation that is intuitively indubitable, but irreducible to the rules of derivation of the LANGUAGE under consideration.

An indirect proof, calling for a less hazardous search, could run as follows. The standard, informal explanation of the relation of L-consequence in terms of truth transmission implies that any logical sentence, i.e., sentence whose symbols are all logical, is L-determinate (it is analytic if it is true, contradictory if false). For, if such a sentence is true (false, respectively), it keeps on being so under all reinterpretations of its extra-logical symbols (since there are none). Thus, it is analytic (contradictory, respectively). Let us take this L-determinacy as a requirement to be met by any explication of the relation of L-consequence for it to be adequate. If L-consequence is explicated in terms of derivability, and, correlatively, analyticity (contradiction, respectively) explicated as demonstrability (refutability, respectively), for such explications to be adequate, every logical sentence should be demonstrable or refutable, i.e., decidable. If this turned out not to be the case, then the explication in question would prove to be inadequate (thus incomplete, if sound).

Before Gödel, logicians did not doubt that this condition was indeed fulfilled in the case of LANGUAGES like that of *Principia Mathematica* or those that Carnap would take as examples in *Syntax*. Nor did they doubt that, if a LANGUAGE proved to be incomplete, it could always be made complete by a well-calculated strengthening of the rules of derivation.[6] This is precisely what Gödel refuted by proving the existence of undecidable logical sentences for a broad class of LANGUAGES, providing they had a certain property stronger than simple consistency, viz, $\omega$-consistency.[7] Among them are not only the LANGUAGES mentioned above, but any $\omega$-consistent LANGUAGE containing a minimum of arithmetic and having rules of transformation meeting the usual requirements of effectiveness. And so, no strengthening of the rules of transformation of such a LANGUAGE by new rules of transformation complying with these requirements would enable one to overcome the inadequacy (thus incompleteness, if not unsoundness) in question.

After Gödel, Carnap could write, in *Syntax*:

> One of the chief tasks of the logical foundations of mathematics is to set up a formal criterion of validity, that is, to state the necessary and sufficient conditions which a sentence must fulfill in order to be valid (correct, true) in the sense understood in classical mathematics. (LSL, p. 100)

But why try to overcome Gödel's theorem? Strangely enough, Carnap does not even ask the question. What I have said so far suggests a change of perspective. If an explication of the relation of L-consequence is adequate, then a non-Gödelian, so to speak, formal criterion of validity for mathematics must follow. No such criterion, no adequacy. Now, not limited to mathematical sentences, the chief task of philosophy remains to be done, viz., set up an adequate explication of the relation of L-consequence.

Before going back to the book that Carnap actually wrote in possession of full knowledge and technical mastery of Gödel's work, let us make a last remark. If an adequate explication of the concept of L-consequence in terms of derivability were possible, and a LANGUAGE were rich enough to contain (be this in a suitably coded form) its elementary SYNTAX (which, as we know well, is perfectly possible and is not even at all exceptional, quite the contrary), then this LANGUAGE would contain, let us not say the totality, but at least the philosophically interesting part of its own SYNTAX (in Carnap's eyes: definitions of predicates of L-consequence, analyticity, etc.). Did Carnap, before Gödel and at least for a time, entertain the idea of *such* a development of the SYNTAX of a LANGUAGE within this LANGUAGE itself and thus the hope of a *radical* refutation of the thesis of the ineffability of SYNTAX defended by Wittgenstein in the *Tractatus*? That is a question for Carnap scholars.[8]

No matter how *philosophically* significant this idea and, therefore, also the proof of its unworkability conducted by Carnap himself in *Syntax* may be, I shall be content here to take note of the *technical* need Carnap had to define the fundamental concepts of the SYNTAX of object-LANGUAGES I and II in more powerful meta-LANGUAGES. The following sections will be devoted to the analysis of the manner in which, recourse to more powerful meta-LANGUAGES being technically understood and therefore accepted, Carnap explicated the relation of L-consequence for object-LANGUAGES I and II while indiscriminately turning away from the concepts of absolute truth and relative truth and, therefore, from the standard, informal explanation in which one or the other (in actual fact, the concept of relative truth) plays a crucial role, and did so without any possibility of falling back on an explication in terms of derivability.

## 3 On Carnap's SYNTACTICAL explication for LANGUAGE I

### 3.1 Reducing LANGUAGE I to its essentials: LANGUAGE $I_0$

The purely logico-mathematical part, taken to be purely logical, of Carnap's LANGUAGE I is a version of the weak form of arithmetic whose idea dates back to Skolem (1923) and which nowadays goes by the name of *primitive recursive arithmetic* (PRA). The essential feature of PRA is the absence of the usual, unlimited quantifiers and, more generally, of any unlimited operator involving the binding of a (numerical) variable. LANGUAGE I itself is obtained from its purely logical part by adding finitely or infinitely[9] many extra-logical symbols, which are *n*-adic ('*n-termed*') predicates or functors, for any natural number $n \neq 0$. Be they logical or extra-logical, symbols of LANGUAGE I may be either primitive, i.e., undefined, or defined from primitive or already defined symbols. If they are defined, they must exclusively be explicitly defined, except for functors, which can *also* be defined in a recursive primitive ('*regressive*') way.[10] Explicitly defined symbols are taken to belong to LANGUAGE I as much as primitive and primitive-recursively defined symbols, but, of course, they can be eliminated.

A striking feature of LANGUAGE I is its richness in primitive symbols and, correlatively, in formation and derivation rules. Carnap notes that certain primitive symbols could be explicitly defined and, thus, eliminated. He mentions symbols of disjunction, conjunction, and equivalence, explicitly definable in terms of negation and implication; limited existential quantifiers ('*operators*'), explicitly definable in terms of negation and limited universal quantifiers ('*operators*'); and a certain operator K, explicitly definable in terms of connectives ('*junction symbols*') and limited, existential and universal quantifiers. Supposing that all these symbols mentioned have been eliminated, Carnap believes that limited universal quantifiers could not in turn be explicitly defined and thus eliminated (LSL, p. 31, l. 16–19). He could, however, be proved wrong.[11]

By carrying out all the eliminations mentioned so far and with some slight modifications in what remains, duly pointed out, below, along the way, a reduced version of LANGUAGE I, call it $I_0$, is obtained whose rules of formation and of transformation (more specifically, derivation) are collectively much simpler, yet as strong as those of LANGUAGE I.

Primitive symbols of LANGUAGE $I_0$ are:

- left and right parentheses, comma: logical symbols for punctuation;
- '$^{\prime}$': logical symbol corresponding to the function of (immediate) succession in the natural number series, without for all that being dubbed a (monadic) functor;
- '0': logical symbol designating 0, without for all that being dubbed a (0-adic) functor;
- '$\sim$' and '$\supset$': logical symbols for negation and implication;
- '=': logical symbol corresponding to relation of numerical identity, without for all that being dubbed a (dyadic) predicate;
- denumerably many (numerical) variables, which are logical symbols;
- for every $n > 0$, denumerably[12] many *n*-adic predicates, which are extra-logical symbols;
- for every $n \geq 0$,[13] denumerably[14] many *n*-adic functors, which are extra-logical symbols.

As for symbols that are not primitive:

- there are no *explicitly* defined symbols;
- finitely many functors (only functors and as many as desired) can be introduced by means of primitive recursive definitions; these functors are logical or extra-logical depending on whether their definition contains only logical symbols or also contains extra-logical symbols.

| **symbols** | **logical** | **extra-logical** |
|---|---|---|
| primitive | '(', ')', 'ˈ', '0', '~', '⊃', '=', variables | all predicates, all primitive functors |
| explicitly defined | ... | ... |
| primitive-recursively defined | all defined functors whose definition contains only logical symbols | all defined functors whose definition contains extra-logical symbols |

In the presentation of rules of formation and of derivation of LANGUAGE $I_0$, I shall use a part of Carnap's gothic symbolism, and I shall do so subsequently, without further explanation. On the whole, in this presentation, this symbolism will amount to the following:

- '$\mathfrak{nu}$' is a SYNTACTICAL constant designating numerical constant '0';
- '$\mathfrak{z}_1$', '$\mathfrak{z}_2$', etc. are SYNTACTICAL variables of (numerical) variable;
- '$\mathfrak{Z}_1$', '$\mathfrak{Z}_2$', etc. are SYNTACTICAL variables of numerical expression;
- '$\mathfrak{pr}_1^n$', '$\mathfrak{pr}_2^n$', etc. are SYNTACTICAL variables of $n$-adic predicate ($n > 0$);
- '$\mathfrak{fu}_1^n$', '$\mathfrak{fu}_2^n$', etc. are SYNTACTICAL variables of $n$-adic functor ($n \geq 0$);
- '$\mathfrak{S}_1$', '$\mathfrak{S}_2$', etc. are SYNTACTICAL variables of sentence;
- '$\mathfrak{K}_1$', '$\mathfrak{K}_2$', etc. are SYNTACTICAL variables of class of (well- or ill-formed) expressions.

Here are formation rules. *Numerical expressions* are variables and constant $\mathfrak{nu}$, and (by induction from this first step) expressions of the form $\mathfrak{Z}_1{}^{\mathsf{I}}$ or $\mathfrak{fu}_1^n(\mathfrak{Z}_1, \mathfrak{Z}_2, \ldots, \mathfrak{Z}_n)$[15]; among numerical expressions are numerals[16] $\mathfrak{nu}$, $\mathfrak{nu}^{\mathsf{I}}$, $\mathfrak{nu}^{\mathsf{II}}$, etc. I shall note $\mathfrak{nu}^{(n)}$ the numeral obtained from $\mathfrak{nu}$ by $n$ accentuations with 'ˈ' ($n \geq 0$). *Sentences* are expressions of the form $\mathfrak{Z}_1 = \mathfrak{Z}_2$ or $\mathfrak{pr}_1^n(\mathfrak{Z}_1, \mathfrak{Z}_2, \ldots, \mathfrak{Z}_n)$, and (by induction from this first step) those of the form $\sim(\mathfrak{S}_1)$ or $(\mathfrak{S}_1) \supset (\mathfrak{S}_2)$. As usual in logic or in mathematics, we do not feel practically compelled not to omit any parenthesis.

A primitive recursive definition of a $n$-adic functor ($n > 0$), $\mathfrak{fu}_1^n$, is presented by Carnap as a system of equations of the form:

$$(a)\ \mathfrak{fu}_1^n(\mathfrak{nu}, \mathfrak{z}_2, \ldots, \mathfrak{z}_n) = \mathfrak{Z}_1,$$

$$(b)\ \mathfrak{fu}_1^n(\mathfrak{z}_1{}^{\mathsf{I}}, \mathfrak{z}_2, \ldots, \mathfrak{z}_n) = \mathfrak{Z}_2,$$

where, 1°) $\mathfrak{z}_1, \mathfrak{z}_2, \ldots, \mathfrak{z}_n$ are distinct variables; 2°) no variable can occur free on the right hand side unless it already occurs free on the left hand side of the same equation; 3°) every occurrence of $\mathfrak{fu}_1^n$ in $\mathfrak{Z}_2$ is followed by $\mathfrak{z}_1, \mathfrak{z}_2, \ldots, \mathfrak{z}_n$. It must be understood that $\mathfrak{fu}_1^n$ does not occur in $\mathfrak{Z}_1$, but does occur in $\mathfrak{Z}_2$, and that the definition in question is relative to other functors occurring in $\mathfrak{Z}_1$ or in $\mathfrak{Z}_2$.

As for derivation rules – viz, axioms ('*primitive sentences*') explicitly or schematically given, and rules of inference – there is nothing original about them, except that the absence of quantifiers, which free variables make up for as far as possible, requires us to formulate the principle of finite induction ('*rule of complete induction*'), not as an axiom schema, but as a primitive, two premise rule of inference (see below). Axioms relative to the connectives '$\sim$' and '$\supset$' are a schematic version of Łukasiewicz's axioms for the propositional calculus (LSL, p. 96); those relative to identity '=' express reflexivity of identity and indiscernibility of identicals; those relative to arithmetical symbols $\mathfrak{nu}$ and '$^{\prime}$' correspond to Peano's third and fourth axioms, which state that 0 is the successor of no number and that distinct numbers have distinct successors. The rules of inference are the rule of substitution, the rule of detachment ('*rule of implication*'), and the rule of finite induction, corresponding to Peano's fifth axiom, which allows inference of a sentence, $\mathfrak{S}_1$, from sentences of the forms $\mathfrak{S}_1(\mathfrak{z}_1/\mathfrak{nu})$ and $\mathfrak{S}_1 \supset \mathfrak{S}_1\ (\mathfrak{z}_1/\mathfrak{z}_1^{\prime})$, where the last two sentences result from substituting '0' and $\mathfrak{z}_1^{\prime}$, respectively, for $\mathfrak{z}_1$ in $\mathfrak{S}_1$.[17]

A derivation of a conclusion from an effective,[18] possibly empty, class of premises is a finite sequence of sentences, $\mathfrak{S}_1, \mathfrak{S}_2, \ldots, \mathfrak{S}_m$, such that: each sentence of the sequence is either an axiom, or one of the two equations of a definition, or one of the premises, or lastly results from sentences which precede it in the sequence by applying a rule of inference; and such that $\mathfrak{S}_m$ is the conclusion in question.

There are no particular axioms ('*primitive sentences*') governing the use of extra-logical symbols (definitional equations are not taken to be such).

## 3.2 Carnap's idea comes from Gödel's proof

Actually, closer scrutiny of Gödel's work would supply Carnap with what constitutes, assuming the hypothesis of (simple) consistency, a *direct* proof of inadequacy (incompleteness, if not unsoundness) of the explication of the relation of L-consequence for the LANGUAGES under consideration in terms of derivability, namely, a rule of transformation, later known as the *rule of infinite induction*, or $\omega$-*rule*, that is intuitively indubitable, at least if the arithmetical symbols are treated as logical symbols, and yet irreducible to (possibly new) rules of derivation. Carnap would not be the first to learn this lesson from Gödel's work. Tarski did so before him in the original version of *Wahrheitsbegriff* (1933a) and in a less well-known article (1933b) devoted to the concepts of $\omega$-consistency and $\omega$-completeness, published the same year. Moreover, he would do so again, after the first edition of *Syntax*, in his article on the concept of logical consequence (1936b). However, Carnap would definitely be the first to try to explicate the relation of logical consequence by taking Gödel's work into account in this way, as Tarski would acknowledge in this latter article (1936b).[19]

Indeed, Gödel would not have been content to show the existence of an undecidable logical sentence in every LANGUAGE of the kind indicated above with an easy proof mobilizing the concept of truth and assuming the soundness, with respect to truth, of the LANGUAGE under consideration.[20] He actually constructed a logical sentence of the form $\forall \mathfrak{z}_1 \mathfrak{S}_1$, where $\mathfrak{S}_1$ is a sentence without quantifiers and $\mathfrak{z}_1$ a variable ranging over the class of natural numbers, which he showed is undecidable in the LANGUAGE under consideration, unless the latter is $\omega$-inconsistent. Now, in the course of the proof, it turns out that this sentence is such that all its numeral instances, $\mathfrak{S}_1(\mathfrak{z}_1/\mathfrak{nu})$, $\mathfrak{S}_1(\mathfrak{z}_1/\mathfrak{nu}^{\mathrm{I}})$, $\mathfrak{S}_1(\mathfrak{z}_1/\mathfrak{nu}^{\mathrm{II}}), \ldots$, are provable, without itself being so, unless the LANGUAGE under consideration is (simply) inconsistent. Then, if one takes the arithmetical symbols (including the variables, supposed to range over natural numbers) for being logical symbols, as Carnap legitimately does, then a universal sentence is intuitively an L-consequence of the class of its numeral instances. In other words, the $\omega$-rule, which enables one to go from the numeral instances of a universal sentence to this sentence itself,

$$\frac{\mathfrak{S}_1(\mathfrak{z}_1/\mathfrak{nu}), \mathfrak{S}_1(\mathfrak{z}_1/\mathfrak{nu}^{\mathrm{I}}), \mathfrak{S}_1(\mathfrak{z}_1/\mathfrak{nu}^{\mathrm{II}}), \ldots}{\forall \mathfrak{z}_1 \mathfrak{S}_1},$$

is intuitively valid. And last, but not least, what emerges through Gödel's work is that, no matter how strong the rules of derivation of the LANGUAGE under consideration may be, they will never have the strength of the $\omega$-rule, and that the latter (and the relation of L-consequence for this LANGUAGE along with it) remains irreducible to rules of derivation, unless this LANGUAGE is (simply) inconsistent.

Substituting 'sentence' for 'universal sentence', and '$\mathfrak{S}_1$' for '$\forall \mathfrak{z}_1 \mathfrak{S}_1$', the latter considerations of the preceding paragraph apply to LANGUAGE $\mathrm{I}_0$.

### 3.3 Carnap's SYNTACTICAL explication

*From now on, up to the end of section 3.6, for the sake of convenience, I shall pretend that Carnap did for* LANGUAGE $\mathrm{I}_0$ *what he in fact did* (mutatis mutandis) *for* LANGUAGE I.

It was this last point of Gödel's work which was going to give Carnap an idea for a third way between the two that, rightly (as concerns the explication in terms of derivation) or wrongly (as concerns the explication inspired by the standard, informal explanation), he considered to be dead ends: explicating the relation of L-consequence for LANGUAGE $I_0$ in terms of the possibility of going from premises to a conclusion by successive applications of certain rules of a new kind. Unlike the rules of derivation, or *d-rules*, which had to be finitary (*'definite'*), the rules now called for, dubbed *rules of L-consequence* (*'rules of consequence'*), or *c-rules*, could be infinitary (*'indefinite'*), as the $\omega$-rule was.

In the case of LANGUAGE $I_0$, one might have expected Carnap to introduce the $\omega$-rule itself, the rule of *infinite* induction, instead of the usual rule of *finite* induction, to strengthen the initial rules, hoping in that way to obtain a sound and complete, in short, an adequate explication of the relation of L-consequence for this LANGUAGE in terms of c-rules. Given a class of premises, the class of its L-consequences would have been the closure of the class of premises under the new system of rules, the so-called c-rules. I shall note $\models_\Omega$ the relation of L-consequence so defined, and $I_\Omega$ the LANGUAGE whose rules of transformation are the c-rules in question.

However, that is not what Carnap did, and it is not obvious that what he did do amounts to the same thing. It seems that Carnap let himself be led by the idea of a close analogy between the two concepts of *deduction*[21] which were to be taken into account, the first being the usual concept of derivation, and the second being the concept of *consequence-series*. For the sake of convenience, I shall outline the analogy (the leading idea of which I attribute to Carnap) in terms of *d-deducibility* for derivability, and *c-deducibility* for the relation of L-consequence between a class of sentences and a sentence.

| **Two kinds of *deduction*** | |
|---|---|
| A d-*deduction* of a conclusion from an effective class of premises . . . | A c-*deduction* of a conclusion from a class of premises . . . |
| . . . is a ***finite*** series of *sentences* such that . . . | . . . must be a ***finite*** series of *classes of sentences* such that . . . |
| . . . 1°) every sentence is a premise or a definition-sentence or is *directly* d-deducible from sentences which precede it in the series by applying a *finitary* rule of transformation (a d-rule), and . . . | . . . 1°) the first class is the class of premises, every other class is *directly* c-deducible from the class which directly precedes it in the series by applying a *possibly infinitary* rule of transformation (a c-rule), and . . . |
| . . . 2°) the last sentence is the conclusion. | . . . 2°) the last class is the singleton of the conclusion. |

The point to be emphasized is the following. Just as every conclusion which is d-deducible from an effective class of premises can be reached in *finitely* many steps from this class, Carnap seems to believe that every conclusion which is c-deducible from a class of premises should be reachable in *finitely* many steps from that class. It could be judged, with good reason, that such a constraint is vacuous, since the steps in question in the case of a c-deduction can be infinitary, and that, in a general way, every finite or transfinite series of possibly infinitary steps can be reduced to a finite series of possibly infinitary steps, and even to a single such step. But once the kind of step authorized is fixed, there is no general reason for the finite or transfinite character of the series not to have an effect upon the extension of what it allows one to reach. I suspect that Carnap missed the problem.

Be that as it may, here is how he precisely defined the relation of L-consequence, that I shall note $\models^{C34}$, between a class of sentences, $\mathfrak{K}_1$, and a sentence, $\mathfrak{S}_1$. In three stages:

(a) $\mathfrak{K}_1 \models^{C34}_{\text{direct}} \mathfrak{K}_2$ if, and only if, every sentence of $\mathfrak{K}_2$ is derivable from a finite subclass of $\mathfrak{K}_1$ without resorting to the rule of finite induction, or can be obtained from an infinite subclass of $\mathfrak{K}_1$ by the $\omega$-rule.

(b) A *consequence-series* is any finite series, $\langle \mathfrak{K}_1, \mathfrak{K}_1, \ldots, \mathfrak{K}_n \rangle$, of finite or infinite classes of sentences such that $\mathfrak{K}_n$ is a singleton and $\mathfrak{K}_1 \models^{C34}_{\text{direct}} \mathfrak{K}_2 \models^{C34}_{\text{direct}} \cdots \models^{C34}_{\text{direct}} \mathfrak{K}_n$.

(c) $\mathfrak{K}_1 \models^{C34} \mathfrak{S}_1$ if, and only if, there exists a consequence-series leading from $\mathfrak{K}_1$ to $\{\mathfrak{S}_1\}$.

### 3.4 Is the relation defined by Carnap closed under the rule of infinite induction? Theorem of closure

I shall also note $\models_\omega$ the relation $\models^{C34}$, and $\mathrm{I}_\omega$ the LANGUAGE whose rules of transformation are the Carnapian c-rules presented just above (section 3.3, *in fine*; compare with the definitions of $\models_\Omega$ and $\mathrm{I}_\Omega$ given in the same section).

It is obvious that the extension of the relation $\models_\omega$ so defined is included in the extension of the relation $\models_\Omega$ considered above (section 3.3), but not that the former is identical to the latter. By definition, the class of L-consequences, in the sense of the relation $\models_\Omega$, of a class of sentences is closed under the $\omega$-rule: $\mathfrak{K}_1$ being any class of sentences, $\mathfrak{S}_1$ any sentence, and $\mathfrak{z}_1$ any variable,

(1) if, for every $n$, $\mathfrak{K}_1 \models_\Omega \mathfrak{S}_1(\mathfrak{z}_1/\mathrm{nu}^{(n)})$, then $\mathfrak{K}_1 \models_\Omega \mathfrak{S}_1$;

but is it the same for the relation $\models_\omega$? The question is thus the following (using the same notation and '¿' for 'Is it the case that'):

(Q1) ¿ if, for every $n$, $\mathfrak{K}_1 \models_\omega \mathfrak{S}_1(\mathfrak{z}_1/\mathrm{nu}^{(n)})$, then $\mathfrak{K}_1 \models_\omega \mathfrak{S}_1$?

Rosser was the first to take an interest in this kind of question. In particular, from Rosser (1937) it results that, if a question analogous to Q1 were asked for LANGUAGE $\mathrm{II}_0$, to be defined later (section 4.1), the answer would be negative. However, for LANGUAGE $\mathrm{I}_0$, the positive answer to Q1 is given by the following theorem, which can easily be obtained as a corollary of the theorem of relative adequacy to be stated in the next section (section 3.5):

THEOREM OF CLOSURE – *The class of L-consequences, in the sense of* $\models_\omega$*, of a class of sentences is closed under the* $\omega$*-rule.*

## 3.5 Is Carnap's explication adequate with respect to Tarski's explication? Theorem of relative adequacy

A more interesting question for us is whether Carnap's explication of the relation of L-consequence for LANGUAGE $\mathrm{I}_0$ is equivalent to the explication which would result from applying, to that same LANGUAGE and for the same division of primitive vocabulary into logical and extra-logical symbols, the general explication given by Tarski in 1936. Noting $\models^{T36}$ the Tarskian explicatum for LANGUAGE $\mathrm{I}_0$, the question would, *grosso modo*, be the following:

(Q2) ¿ the two explicata, $\models^{C34}$ and $\models^{T36}$, are coextensive?

However, let us be more accurate. All variables occurring in sentences of LANGUAGE $\mathrm{I}_0$ are free, and an open sentence of this LANGUAGE not in a position of a subformula must be understood as its universal closure would be. This universal closure belongs, not to LANGUAGE $\mathrm{I}_0$, but to the extension of this LANGUAGE obtained by adjoining to it the universal quantifier, parsed as a logical symbol and understood in a classical way, and rules to govern its use in accordance with this understanding.

Let $\mathfrak{K}_1$ be a class of sentences, and $\mathfrak{S}_1$ a sentence, of LANGUAGE $\mathrm{I}_0$. Let us note $\mathfrak{K}_{\mathfrak{ddf}}$ the class of extra-logical (or *d*escriptive) *def*inition-sentences of this LANGUAGE; $(\forall)\mathfrak{K}_{\mathfrak{ddf}}$ and $(\forall)\mathfrak{K}_1$, the classes of universal closures of sentences of $\mathfrak{K}_{\mathfrak{ddf}}$ and $\mathfrak{K}_1$, respectively; $(\forall)\mathfrak{S}_1$ the universal closure of the sentence $\mathfrak{S}_1$. The following theorem gives the best positive answer to Q2 that we could reasonably hope to obtain:

THEOREM OF RELATIVE ADEQUACY – *The universal closures of the L-consequences, in the sense of* $\models^{C34}$*, of a class* $\mathfrak{K}_1$ *of sentences are logical consequences, in the sense of* $\models^{T36}$*, of the union of the class* $(\forall)\mathfrak{K}_{\mathfrak{ddf}}$ *of the universal closures of the extra-logical definitional sentences and of the class* $(\forall)\mathfrak{K}_1$ *of universal closures of the sentences of the class* $\mathfrak{K}_1$ (theorem of relative soundness) *and conversely* (theorem of relative completeness). *In other words,*

$$\mathfrak{K}_1 \models^{C34} \mathfrak{S}_1 \textit{ if, and only if, } (\forall)\mathfrak{K}_{\mathfrak{ddf}} \cup (\forall)\mathfrak{K}_1 \models^{T36} (\forall)\mathfrak{S}_1.$$ [22]

### 3.6 Is Carnap's explication absolutely adequate?

As a matter of principle, it is obviously impossible to prove formally the adequacy of a formal explication of an informal concept (Carnap 1950b, ch. 1). In the best case, it is only possible to obtain theorems of relative adequacy. For want of having the idea of any such theorem for LANGUAGE $I_0$, Carnap had to be content with proving the theorem of L-determination of logical sentences of LANGUAGE $I_0$ (LSL, p. 40). Our theorem of relative adequacy is obviously much stronger, and it would be as good as a proof of absolute adequacy for anybody who considered Tarski's explication absolutely adequate. Even so, however, Carnap's explication would not be exempt from any criticism.

Tarski would criticize it for not being generalizable.[23] I shall show (section 4.3) what is superficial and unjustified about that criticism. If Carnap is to be reproached in any way for his explication in the case of LANGUAGE $I_0$ (and this reproach would hold for its generalizations), it would not be for its lack of generality, but for its lack of naturalness, or of self-evidence, qualities possessed by the standard, informal explanation in terms of transmission of truth (*relative to* every interpretation), once this latter concept is explicated. Carnap was not in a position to admit the standard, informal explanation because of what he saw as being the irreducibly non-SYNTACTICAL nature of the concept of truth involved in that explanation. Only Tarski, having removed this hindrance beforehand, could do that.

When one is familiar with the explication that Tarski would give of the relation of logical consequence in 1936, after he had read *Syntax*, by simply taking up the standard, informal explanation again on the basis of the explication of the concept of truth that he would have given previously, one cannot help but be persuaded of its superiority over the one Carnap proposed for LANGUAGE $I_0$. Carnap's explication for LANGUAGE $I_0$ is not, though, void of interest in that it pinpoints the origin of the difference between the relation of derivability and that of logical consequence for this LANGUAGE as lying in a difference between the rule of finite induction and the rule of infinite induction (or more exactly a certain use of the latter rule, which, according to the theorem of closure, turns out to be equivalent to the normal use of that rule).

One last remark will be in order, just to remind the reader of the method adopted in the present section 3. In section 3.1, for the sake of simplicity, I replaced LANGUAGE I by LANGUAGE $I_0$, but, because they are in fact equivalent in expressive richness and derivative strength, every claim made about the latter from section 3.2 on could have been made, *mutatis mutandis*, about the former. In particular, the theorems of closure and relative adequacy hold for both.

## 4 On Carnap's SYNTACTICAL explication for LANGUAGE II

### 4.1 Reducing LANGUAGE II to its essentials: LANGUAGE $\text{II}_0$

LANGUAGE II is what could be called an applied arithmetic of order $\omega$, or an applied theory of types whose individuals are natural numbers and which includes second-order Peano arithmetic. Even more than LANGUAGE I, LANGUAGE II is remarkable for its primitive, expressive richness. As I did for LANGUAGE I, and for the same reasons, I shall propose a simplified version, noted $\text{II}_0$, of LANGUAGE II.

LANGUAGE $\text{II}_0$ is obtained from LANGUAGE II essentially by eliminating the following symbols (and everything which, in the rules of formation and the rules of transformation, more specifically of derivation, governs their use): all defined symbols of LANGUAGE II; variables replaceable by sentences; symbols of disjunction, conjunction, and equivalence; limited and unlimited K-operators; limited and unlimited, existential quantifiers; and limited, universal quantifiers. There are other, minor, modifications, notably, the introduction of infinitely many primitive, extra-logical, numerical symbols. Here is a rapid presentation of LANGUAGE $\text{II}_0$.

First, *types* are defined as being of one of the following forms:

- 0; then (by induction from this first step)
- $t_1$, $t_2$ (*sic*), where $t_1$ and $t_2$ are types;[24]
- $(t_1)$, where $t_1$ is a type;
- $(t_1 : t_2)$, where $t_1$ and $t_2$ are types.

For the sake of convenience, I shall qualify types of form 0, or $(t_1)$, or $(t_1\text{: } t_2)$ as *standard* and those of the form $t_1$, $t_2$ as *non-standard*. Every type is analysable in one, and only one, way as being of form $t_1, t_2, \ldots, t_n$, where $n \geq 1$ and types $t_1, t_2, \ldots, t_n$ are standard. I shall qualify that form as *canonical* (notice that a standard type always appears in canonical form). Identity between two non-standard types of canonical forms $t_1, t_2, \ldots, t_p$, and $t_{p+1}, t_{p+2}, \ldots, t_{p+q}$, respectively, implies that $p = q$ and must naturally be understood as the conjunction of identities between the standard types $t_i$ and $t_{p+i}$ for $1 \leq i \leq q$.[25]

*Logical, primitive symbols* are left and right parentheses, comma, '|', '0', '$\sim$' and '$\supset$', '=', and, for every standard type, denumerably many *variables* of this type. '0' is a *numerical constant* of type 0. Variables are *numerical, predicate-* or *functor-variables*, depending on whether their type is of the form 0, $(t_1)$, or $(t_1 : t_2)$, respectively.

*Extra-logical, primitive symbols* are, for every standard type, denumerably many *constants* of this type: *numerical* constants other than '0',[26] *predicate-* or *functor*-constants, according to whether their type is of form 0, $(t_1)$, or $(t_1 \colon t_2)$, respectively.

There are *no defined symbols.*

| symbols | logical | extra-logical |
|---|---|---|
| primitive | '(', ')', ',', '0' (numerical constant, of type 0), '$\sim$', '$\supset$', '=', variables of every standard type, '$\forall$' | numerical constants (of type 0) other than '0', predicate-constants and functor-constants of every standard type $\neq 0$ |
| defined | ... | ... |

*The class of well-formed expressions* and certain parts of it could be inductively defined in a more usual way, but let us adopt Carnap's style of presentation. To begin with, let us introduce part of his gothic symbolism.

- '$\mathfrak{nu}$' is a SYNTACTICAL constant designating the *numerical constant* '0';
- '$\mathfrak{v}_1$', '$\mathfrak{v}_2$', etc. are SYNTACTICAL variables of *variable* of any standard type;
- '$\mathfrak{Z}_1$', '$\mathfrak{Z}_1$', etc. are SYNTACTICAL variables of *numerical expression* of type 0;
- '$\mathfrak{Pr}^n_1$', '$\mathfrak{Pr}^n_2$', etc. ('$\mathfrak{pr}^n_1$', '$\mathfrak{pr}^n_2$', etc., respectively) are SYNTACTICAL variables of *predicate-expression* (*predicate-symbol*, respectively) of type of the form $(t_1)$, with $t_1$ of the canonical form $t_2, t_3, \ldots, t_{n+1}$;[27]
- '$\mathfrak{Fu}^n_1$', '$\mathfrak{Fu}^n_2$', etc. ('$\mathfrak{fu}^n_1$', '$\mathfrak{fu}^n_2$', etc., respectively) are SYNTACTICAL variables of *functor-expression* (*functor-symbol*, respectively) of type of the form $(t_1 \colon t_2)$, with $t_1$ of the canonical form $t_3, t_4, \ldots, t_{n+2}$;[28]
- '$\mathfrak{Arg}^n_1$', '$\mathfrak{Arg}^n_2$', etc. are SYNTACTICAL variables of *n-termed argument-expression* or *n-termed value-expression*, whose type is of the canonical form $t_1, t_2, \ldots, t_n$;
- '$\mathfrak{S}_1$', '$\mathfrak{S}_2$', etc. are SYNTACTICAL variables of *sentence* (no type);
- '$\mathfrak{A}_1$', '$\mathfrak{A}_2$', etc. are SYNTACTICAL variables of (well- or ill-formed) expression;
- '$\mathfrak{K}_1$', '$\mathfrak{K}_2$', etc. are SYNTACTICAL variables of class of expressions.

Second, let us define in an overall inductive way, numerical expressions (1), predicate-expressions (2), functor-expressions (3), argument- (or value-) expressions (4), and let us systematically assign them a type.

(1) *Numerical expressions* are symbols of type 0, or are of the form $\mathfrak{Z}_1{}^1$ or $\mathfrak{Fu}^n_1(\mathfrak{Arg}^n_1)$, with $\mathfrak{Arg}^n_1$ of type $t_1$ and $\mathfrak{Fu}^n_1$ of type $(t_1 : 0)$; $\mathfrak{Z}_1{}^1$ is of type 0 and, under these conditions, $\mathfrak{Fu}^n_1(\mathfrak{Arg}_1)$ is of type 0.

(2) *Predicate-expressions* are symbols of type of form $(t_1)$, or are of the form $\mathfrak{Fu}^n_1(\mathfrak{Arg}^n_1)$, with $\mathfrak{Arg}^n_1$ of type $t_1$, and $\mathfrak{Fu}^n_1$ of type of form $(t_1 : (t_2))$; under these conditions, $\mathfrak{Fu}^n_1(\mathfrak{Arg}^n_1)$ is of type $(t_2)$. *Predicate-symbols* (*-variables*, *-constants*, respectively) are predicate-expressions consisting of a single symbol (variable, constant, respectively).

(3) *Functor-expressions* are symbols of type of form $(t_1\colon t_2)$, or are of the form $\mathfrak{Fu}^n_1(\mathfrak{Arg}^n_1)$, with $\mathfrak{Arg}^n_1$ of type $t_1$ and $\mathfrak{Fu}^n_1$ of type of form $(t_1\colon (t_2 : t_3))$; under these conditions, $\mathfrak{Fu}^n_1(\mathfrak{Arg}^n_1)$ is of type $(t_2 : t_3)$. *Functor-symbols* (*-variables*, *-constants*, respectively) are functor-expressions consisting of a single symbol (variable, constant, respectively).

(4) An *n-termed argument-* or *n-termed value-expression* is (a *single* expression) of the form $\mathfrak{A}_1, \mathfrak{A}_2, \ldots, \mathfrak{A}_n$, where $n \geq 1$ and $\mathfrak{A}_1, \mathfrak{A}_2, \ldots, \mathfrak{A}_n$ are well-formed expressions (separated from one another by commas) of standard types; if these types are $t_1, t_2, \ldots, t_n$, respectively, then the argument- or value-expression in question is of the (*single*) type $t_1, t_2, \ldots, t_n$.

(5) *Sentences* are of one of the following forms: $\mathfrak{A}_1 = \mathfrak{A}_2$, with $\mathfrak{A}_1$ and $\mathfrak{A}_2$ being well-formed and of the same type[29]; $\sim\mathfrak{S}_1$; $\mathfrak{S}_1 \supset \mathfrak{S}_2$; $\forall\mathfrak{v}_1(\mathfrak{S}_1)$; $\mathfrak{Pr}^n_1(\mathfrak{Arg}^n_1)$, with $\mathfrak{Arg}^n_1$ of type $t_1$ and $\mathfrak{Pr}^n_1$ of type $(t_1)$.

It will be noticed that a variable cannot be bound in any well-formed expression of any category other than that of sentences.

A system of *derivation rules*, alias *d-rules* (i.e., *axioms* and *rules of inference*) for LANGUAGE $\text{II}_0$ could be easily drawn from the one that Carnap devised for LANGUAGE II. This system would contain: first, the usual axioms and rules of inference of the simple theory of types, *plus* the axiom schemas of extensionality and of choice, and *minus* the axiom of infinity; second, the Peano axioms for second-order arithmetic, with respect to which an axiom of infinity would be redundant. However, an axiom schema not having a counterpart in Carnap's system (see above, n. 29) should be added to govern the use of the identity symbol between expressions of the same non-standard type. This axiom schema should imply that, if $\mathfrak{A}_1$ and $\mathfrak{A}_2$ are well-formed expressions of the forms $\mathfrak{A}_{1,1}, \mathfrak{A}_{1,2}, \ldots, \mathfrak{A}_{1,n}$ and $\mathfrak{A}_{2,1}, \mathfrak{A}_{2,2}, \ldots, \mathfrak{A}_{2,n}$, respectively, where the latter $2n$ expressions are of standard type, then the identity $\mathfrak{A}_1 = \mathfrak{A}_2$ is equivalent to the conjunction of the $n$ identities $\mathfrak{A}_{1,i} = \mathfrak{A}_{2,i}$, for $1 \leq i \leq n$.

## 4.2 Carnap's explication is uselessly cumbersome

In the case of LANGUAGE II, in sections 34a–f of *Syntax* (LSL, pp. 98–120), the d-rules left the stage open for a horribly and uselessly complicated system of *rules of L-consequence* ('*rules of consequence*'), alias *c-rules*, to which the theorem of L-determination of the logical formulas of LANGUAGE II

(LSL, p. 116) corresponds. Here is what Tarski would very rightly say in his 1936 article of Carnap's attempt:

> [T]his attempt is connected rather closely with the particular properties of the formalized language which was chosen as the subject of investigation. The definition proposed by Carnap can be formulated as follows:
>
> *The sentence X* follows logically *from the sentences of the class K if and only if the class consisting of all the sentences of K and of the negation of X is contradictory.*
>
> The decisive element of the above definition obviously is the concept 'contradictory'. Carnap's definition of this concept is too complicated and special to be reproduced here without long and troublesome explanations.[30] (1936b/1956, pp. 414–15)

It is impossible to read the relevant passage of *Syntax* referred to above and comprehend it without wanting to look for an alternative to Carnap's explication that is provably equivalent in the SYNTAX of LANGUAGE II, but the idea of which would be easier to grasp. The situation would not be essentially different if we transposed Carnap's explication for LANGUAGE II to an explication for LANGUAGE $II_0$.

I know of at least three such alternatives: a first one, of my own devising, inspired by what Carnap did for LANGUAGE I, see section 4.3 below; a second one, Kleene's of 1939, grounded in an inductive definition of the concept of analyticity, see section 4.4 below; a third, Tarski's of 1936, an appropriate rehabilitation of the standard, informal explanation made possible by the prior explication of the concept of truth, see section 4.5 below.

If I am not mistaken, it would be a matter, not of virtuosity, but only of endurance to prove these alternatives to be equivalent one to another, and equivalent to the Carnapian explication.

## 4.3 Alternative I: an explication analogous to Carnap's for LANGUAGE I would have been possible after all

Just after having proposed his explication for LANGUAGE II, Carnap stated that he could have proceeded in a completely different way, analogously to the way he had for LANGUAGE I:

> 'Derivable' is defined as a finite chain of the relation 'directly derivable'. [L-c]onsequence might be analogously [with respect to the definition for Language I] defined as a chain of a simpler relation 'direct [L-]consequence'. [...] In this way the definitions for Language I were previously formulated. In the case of the definitions just given [for Language II] we took a different course. (LSL, pp. 117–18)

In his 1936 article, Tarski declared, on the contrary, that the explication proposed by Carnap for LANGUAGE I was not open to generalization. He even criticized it for precisely that:

> This definition [...] cannot be applied to languages of a more complicated logical structure. (Tarski 1936b/1956, p. 413, n. 3, already quoted)

Who was right, Carnap or Tarski? The question is whether the Carnapian definition, or explication, for LANGUAGE I can, *mutatis mutandis* – or let us say, to link up with Carnap's suggestion, *analogically* – be applied to LANGUAGE II. My answer is that it can. Carnap could very well have, *pace* Tarski, proposed an explication for LANGUAGE II analogous to the one he had proposed for LANGUAGE I. That explication would have been no better or no worse when it came to generality than Tarski's and would have proved to have the same kind of interest as the one I noted above (section 3.6) in the case of LANGUAGE I. I outline the proof of that below, of course substituting LANGUAGES $I_0$ and $II_0$ for LANGUAGES I and II, respectively.

The alternative I propose to outline here essentially consists in deleting the rule of finite induction of LANGUAGE $II_0$ and complementing the remaining rules of transformation with new rules, viz., infinitary rules analogous to the $\omega$-rule, thus obtaining an explication of the relation of L-consequence modelled upon the one given for LANGUAGE $I_0$. The latter explication, remember, corresponded, by definition, to the transformation rules of LANGUAGE $I_\omega$. To express myself in more exact terms than I have above, I must say that it is rather upon the explication corresponding to that of LANGUAGE $I_\Omega$ that I want to model my explication. In other words, I shall model my explication for LANGUAGE $II_0$ upon the one given for LANGUAGE $I_0$, in which the rule of finite induction was replaced from the outset by the *full* $\omega$-rule. I make this choice for the sake, not only of simplicity, but also of caution, in view of Rosser's result mentioned in section 3.4.

Let us generalize the idea of *rule of infinite induction* already contemplated under the name of $\omega$-rule in connection with LANGUAGE $I_0$, to *any standard type whatever*. For type 0, there is no problem. The rule of infinite induction is just the $\omega$-rule for LANGUAGE $II_0$. But, for any standard type different from 0, there is obviously a trap to be avoided. Let us consider, for example, type (0). The rule of infinite induction for type (0) should allow us to infer a sentence of the form $\forall\mathfrak{v}_1(\mathfrak{S}_1)$, with $\mathfrak{v}_1$ being a variable of type (0), from a certain class of instances, let us say the class of all the *canonical* instances (in a sense to be determined), of the sentence $\mathfrak{S}_1$ relative to $\mathfrak{v}_1$, just as the $\omega$-rule would enable us to infer the universal closure of a sentence relative to a numerical variable from the class of all the *numeral* instances of that sentence relative to that variable. The trap to avoid is the one into which we would fall if we admitted as a rule of infinite induction for type (0) that from the class of canonical instances, defined as being certain (to be specified) instances *belonging to* LANGUAGE $II_0$, of $\mathfrak{S}_1$ relative to $\mathfrak{v}_1$, one could infer $\forall\mathfrak{v}_1(\mathfrak{S}_1)$. No matter how the concept of canonical instance could be specified afterwards (including, therefore, if *every* instance of $\mathfrak{S}_1$ *in* LANGUAGE $II_0$ were considered canonical), such a rule would certainly not perform the services expected of it, or, to put it more naïvely and less pragmatically, such a rule would certainly not be intuitively valid. The reason is that 'there are *numerical properties which are not definable* [in LANGUAGE II (or $II_0$)]' (LSL, pp. 106–7).[31] Obviously, what I have just explained for type (0) holds for any standard type different from 0.

The solution consists in making a detour through the LANGUAGE, $II_0^*$, obtained from LANGUAGE $II_0$ by adding certain new constants, which I shall call *canonical*, for every standard type different from 0, and rules to govern their use. For type (0) [(0, 0), etc., respectively], the canonical constants are to be introduced in one-to-one correspondence with the classes of numerals (*sic*)[32] [dyadic relations in extension between numerals, etc., respectively]. All rules of derivation for LANGUAGE $II_0$ are to be extended as rules of transformation of LANGUAGE $II_0^*$.

In order to define the relation of L-consequence for LANGUAGE $II_0^*$: First, replace the principle of finite induction by the ω-rule. Second, for each standard type different from 0, add two rules of transformation to govern the use of the new, canonical constants of that type modelled upon the following rules for type (0): using '$\mathfrak{pr}_1^{1*}$', '$\mathfrak{pr}_2^{1*}$', etc. as SYNTACTICAL variables of canonical constant of type (0), the first rule will assure that if $\mathfrak{nu}^{(n)}$ is a numeral belonging to the class corresponding to $\mathfrak{pr}_1^{1*}$, then $\mathfrak{pr}_1^{1*}(\mathfrak{nu}^{(n)})$ is an axiom; the second rule will be a rule of infinite induction for that type, assuring that from the class of *canonical* instances, $\mathfrak{S}_1(\mathfrak{v}_1/\mathfrak{pr}_1^{1*})$, of a sentence, $\mathfrak{S}_1$, relatively to a variable, $\mathfrak{v}_1$, of type (0), one can infer $\forall\mathfrak{v}_1\mathfrak{S}_1$. Third, and lastly, define a sentence, $\mathfrak{S}_1$, as being an L-consequence of a class, $\mathfrak{K}_1$, of sentences if, and only if, $\mathfrak{S}_1$ belongs to the closure of $\mathfrak{K}_1$ under the aforesaid rules of transformation (including the rules of transformation of LANGUAGE $II_0^*$ other than the rule of finite induction).

It only remains to define the relation of L-consequence for LANGUAGE $II_0$ by restricting to it the relation of L-consequence defined for LANGUAGE $II_0^*$.

## 4.4 Alternative II: Kleene's dramatic simplification of Carnap's explication

Kleene set out a version of his alternative in half a page in his review of *Syntax* (Kleene 1939a, pp. 83–4). He presented it as 'a form of definition of "analytic" and "contradictory" which is more straightforward than Carnap's' (ibid., p. 83). He began by ridding LANGUAGE II of a certain number of elements that contribute nothing to its expressive or derivative power and would uselessly complicate the definition being sought, as I myself did to obtain LANGUAGE $II_0$, but he went one step further in replacing open sentences by their universal closures, so that every sentence be closed.

Kleene's explication presupposes the concept of *valuation*, borrowed from *Syntax* (LSL, pp. 106–10). Here is a brief outline of the relevant definitions. First, the concept of *valuation* of any type is defined by induction on the complexity of types: valuations of type 0 are numerals; valuations of type (0) are classes of numerals; valuations of type (0 : 0) are applications of the class of numerals in itself; etc.[33] Second, the possible *valuations for* a typed symbol other than $\mathfrak{nu}$ are defined as being the valuations of the type of that symbol. Third, the valuation for a typed expression *on the basis of valuations for its typed symbols other than* $\mathfrak{nu}$ is defined in a quite natural way, by induction on the complexity of the expression in such a way that the valuation in question is always of the same type as the expression: $\mathfrak{nu}$ is the unique valuation for itself; if the valuation of $\mathfrak{Z}_1$ on the basis of valuations of its typed symbols other than $\mathfrak{nu}$ is $\mathfrak{nu}^{(n)}$, then the valuation of $\mathfrak{Z}^{|}_1$ on the same basis is $\mathfrak{nu}^{(n+1)}$; if, on the basis of valuations of typed symbols other than $\mathfrak{nu}$ of $\mathfrak{Fu}^1_1(\mathfrak{Z}_1)$, the valuation of $\mathfrak{Fu}^1_1$ is a certain application of the class of numerals in the class of classes of numerals, and the valuation of $\mathfrak{Z}_1$ is a certain numeral, then the valuation of $\mathfrak{Fu}^1_1(\mathfrak{Z}_1)$ on the same basis is the value of that application on this numeral, namely a certain class of numerals; etc.

In the definition quoted below, '$\boldsymbol{B}$', '$\boldsymbol{B}_1$', '$\boldsymbol{B}_2$'[34] are SYNTACTICAL variables of class of valuations for any class of symbols, with one and only one valuation for each symbol; $\boldsymbol{B}_1\boldsymbol{B}_2$ is $\boldsymbol{B}_1 \cup \boldsymbol{B}_2$ (the union of $\boldsymbol{B}_1$ and $\boldsymbol{B}_2$). Kleene defines the concept of analyticity (contradiction, respectively), abridged by 'A' ('C', respectively) in two steps. In the first step, he defines this concept in terms of a relative version of it abridged by 'A—$\boldsymbol{B}$' ('C—$\boldsymbol{B}$', respectively); in the second, he defines this relative version. He does so in a way that displays fascinating concision.

> A sentence [a class of sentences] is A(C), if, for every set $\boldsymbol{B}_1$ of valuations for the descriptive symbols, if any, which occur in the sentence [the sentences of the class], the sentence [every (some) sentence of the class] is A—$\boldsymbol{B}_1$ (C—$\boldsymbol{B}_1$).
>
> $\forall\mathfrak{v}_1\mathfrak{S}_1$ is A—$\boldsymbol{B}_1$ (C—$\boldsymbol{B}_1$) if, for every (some) valuation $\boldsymbol{B}_2$ for $\mathfrak{v}_1$, $\mathfrak{S}_1$ is A—$\boldsymbol{B}_1\boldsymbol{B}_2$ (C—$\boldsymbol{B}_1\boldsymbol{B}_2$). $\sim\mathfrak{S}_1$ is A—$\boldsymbol{B}$ (C—$\boldsymbol{B}$), if $\mathfrak{S}_1$ is not A—$\boldsymbol{B}$ (C—$\boldsymbol{B}$) [alternatively, if $\mathfrak{S}_1$ is C—$\boldsymbol{B}$ (A—$\boldsymbol{B}$)]. $\mathfrak{S}_1 \supset \mathfrak{S}_2$ is A—$\boldsymbol{B}$ (C—$\boldsymbol{B}$), if $\mathfrak{S}_1$ is C—$\boldsymbol{B}$ (A—$\boldsymbol{B}$) or (and) $\mathfrak{S}_2$ is A—$\boldsymbol{B}$ (C—$\boldsymbol{B}$).[35] $\mathfrak{Pr}_1(\mathfrak{Arg}_1)$ is A—$\boldsymbol{B}$ (C—$\boldsymbol{B}$), if the valuation of $\mathfrak{Pr}_1$ on the basis of $\boldsymbol{B}$ does (does not) contain the valuation of $\mathfrak{Arg}_1$ on the basis of $\boldsymbol{B}$. $\mathfrak{Arg}_1 = \mathfrak{Arg}_2$ is A—$\boldsymbol{B}$ (C—$\boldsymbol{B}$), if the valuations of $\mathfrak{Arg}_1$ and $\mathfrak{Arg}_2$ on the basis of $\boldsymbol{B}$ are the same (different). (Kleene 1939a, p. 84)

Kleene does not define the relation of L-consequence between a class of closed *or open* sentences and a closed *or open* sentence (for us, the relation of L-consequence for LANGUAGE $II_0$). However, the way to do that is straightforward. First, as far as only closed sentences are concerned, $\mathfrak{S}_1$ *is a L-consequence of* $\mathfrak{K}_1$ *if, and only if,* $\mathfrak{K}_1 \cup \{\sim\mathfrak{S}_1\}$ *is contradictory*; second, if open sentences are reinstated, $\mathfrak{S}_1$ *is a L-consequence of* $\mathfrak{K}_1$ *if, and only if,* $(\forall)\mathfrak{S}_1$ *is a L-consequence of* $(\forall)\mathfrak{K}_1$.[36]

### 4.5 Alternative III: Tarski's scientific vindication of the standard, informal explanation

What the standard, informal explanation lacked for it to be a genuine explication and acceptable in Carnap's eyes in 1934 was the explication of the concept of truth (in fact, and more precisely, of the concept of relative truth). At the time he was reading Carnap's *Syntax*, Tarski had in fact already nearly supplied the missing explication (indeed, he had only supplied the explication of the concept of absolute truth). To provide his own explication of the relation of logical consequence in 1936 in counterpoint to *Syntax*, it was enough for him to retrieve the standard, informal explanation word for word. In a few luminous pages, the technical nature of which was reduced to a minimum, the matter was settled. Tarski's explication, which does not mobilize the concept of absolute truth, but only that of relative truth, could be made as clearly *SYNTACTICAL*[37] as the preceding alternatives. Carnap would display incredible offhandedness in his 'Intellectual Autobiography' when merely mentioning Tarski's article as an 'interesting paper on semantics' (1963a, p. 61, n. 11). It is not so much the new terminology ('semantics') which bothers me, even though there would be much to say about it,[38] as it is the compliment paid.

Carnap's explication of the relation of L-consequence for LANGUAGE $\text{II}_0$ and alternatives I and II are equivalent to Tarski's. The four explications are simultaneously adequate or inadequate, even though Tarski's has much more in its favour than the others.

Naturally, a point remains to be clarified in Tarski's explication of the concept of truth. It is a matter of the distinction it presupposes between logical and extra-logical terms. In 1936, Tarski was sceptical about the objective well-foundedness of the distinction. Later, he would contemplate grounding this distinction in considerations inspired by Felix Klein's Erlangen programme for geometry.[39]

## 5 Epilogue: Carnap's reaction to Tarski's work and its influence on contemporary philosophy

We know that Carnap soon realized that Tarski's explication of the concept of truth accorded this concept, of ill-repute up until then in the eyes of the members the Vienna Circle, undeniable scientific respectability. By 1939, he would abandon the exclusively SYNTACTICAL point of view to which he had adhered up until then and would adopt the so-called 'semantical' point of view. Without even attempting to sketch the story of this turn and its consequences, I just want to point out at least three reasons why the story would be difficult to tell. The first is the semantical shift of the key words 'syntax' and 'semantics'.[40] The second is that Carnap made the 'semantical' turn by choosing, from among several possible courses recognized as being technically equivalent, the one that consisted in an inversion of the relationship that Tarski had recognized between truth and meaning. The 'semantics' nowadays called vericonditional, of which Davidson is the most prominent representative, is the direct heir of Carnap's inversion. It is no longer the explication of truth that mobilizes the concept of meaning (or a related concept, like that of translation for Tarski), it is inversely the explication of meaning (or a related concept, like that of interpretation for Davidson) that mobilizes the concept of truth. Dummett, and others following in his wake, have believed that they could trace vericonditional 'semantics' back to Frege. It would be more advisable to recognize its true birth in the semantics of early Wittgenstein and that of Carnap. Be that as it may, through Carnap's appropriation of Tarski's 'semantics', the analysis of the relation of logical consequence would undergo the repercussions of the inversion in question. The third and last reason is the change that 'semantics' underwent in the fifties when, following the lead of Tarski and other first rate logicians, it adopted the new, well-known, model-theoretical style that it still has nowadays.[41]

## Notes

1. It is to call these facts to mind that, in this chapter, I have systematically put 'language' and 'syntax' and their derivatives in SMALL CAPITALS.
2. The contrast between 'explication' and 'explanation' here is borrowed from Carnap: 'By a procedure of *explication* we mean the transformation of an inexact, prescientific concept, the *explicandum*, into a new exact concept, the *explicatum*. Although the explicandum cannot be given in exact terms, it should be made as clear as possible by informal explanations and examples' (Carnap 1950b, ch. 1, 'On Explication', summary of §2, p. 3). As the latter sentence makes clear enough, for Carnap at that later time (1950), an explanation of a concept would only be a first step in the process of explication of that concept, meant to specify informally what is to be explicated. In order to help the reader to avoid any confusion, instead of 'explanation', I shall systematically use 'informal explanation'.
3. '[S]emantics becomes a part of the morphology [in other words, of the syntax] of [the object-]language' (Tarski 1936a/1956, p. 406; see also Tarski 1933a/1956, p. 273). 'Tarski has emphasized especially the possibility of finding, for a given formalized language, a purely syntactical property of the well-formed formulas which coincides in extension with the property of being a true sentence' (Church 1956, p. 65).
4. It is this condition that exempts us in what follows (and exempted Carnap in the study of LANGUAGES I and II) from accompanying the term 'derivable' and the related terms by the prefix 'L-'.
5. This condition is banal for transformation rules of formal systems *à la* Frege–Hilbert, and Carnap imposes it on rules of derivation, but not on what he calls 'rules of consequence' (to be understood as 'rules of L-consequence').

6. Its being a matter of logical positivists like Carnap, in his review of *Syntax*, Mac Lane stated: 'Logical positivists formerly distinguished between logic (including mathematics) and empirical science, on the ground that the sentences of logic [i.e. the logical sentences] are resoluble (either demonstrable or refutable), while sentences of [empirical] science need not be resoluble (on the basis of logical rules)' (1938, p. 172, additions between square brackets are my own).
7. J. B. Rosser would later show that the Gödelian hypothesis of $\omega$-consistency can be replaced by the hypothesis of simple consistency (Rosser 1936).
8. There is no evidence for this in the work Carnap published during his lifetime. As for the 'monolinguistic project', as Coffa puts it (1991, pp. 273–8), of Carnap's posthumous *Untersuchungen zur Allgemeinen Axiomatik*, written about 1929 (Carnap 2000), the idea is quite different. For more on this question, see, e.g., Awodey and Carus (2007).
9. I am uncertain about the finite or infinite character of the number of primitive *n*-adic predicates (functors, respectively), for any *n*, in LANGUAGE I.
10. What a definition of the latter kind consists of is recalled below.
11. For a proof, see the unabridged French version of this article (Rouilhan 2008).
12. Carnap is not this explicit, see n. 9.
13. Carnap does not countenance 0-adic functors; I do.
14. Carnap is not this explicit, see n. 9.
15. If $n=0$, $\mathfrak{fu}_1^n\ (\mathfrak{z}_1, \mathfrak{z}_2, \ldots, \mathfrak{z}_n)$ amounts, of course, to $\mathfrak{fu}_1^0$.
16. I am using this term in the sense usual nowadays, which is not Carnap's sense.
17. My notation for substitution differs slightly from Carnap's.
18. Carnap requires that the class of premises be finite (LSL, p. 28). I just require that it be (finite or infinite, but) effective.
19. 'The first attempt to formulate a precise definition of the proper concept of consequence [proper, i.e., not presumed to be reducible to the concept of derivability, PR] was that of Carnap' (and here Tarski refers to *Syntax* [original ed., 1934d] and to the article 'Ein Gültigkeitskriterium....' [1935d] which would be integrated, in a slightly modified form, into the English version [1937, §§34 a–i]), see Tarski (1936b/1956, p. 413). Carnap claims this precedence in *Syntax*: '[T]he term "consequence" [...] has not been defined [for] the languages in use hitherto' (LSL, p. 167, l. 8–9, between parentheses; I have substituted 'for' for the mistranslation 'in').
20. A LANGUAGE of the kind indicated above contains (a possibly coded version of) its own concept of provability, but not, barring inconsistency, its own concept of truth. If this LANGUAGE is sound, therefore, there exists a formula which is true (thus, irrefutable), but not provable.
21. I am here using the term 'deduction' for the first time in the same sense Carnap did on pages 39 and 100 of LSL. This term does not appear anywhere else in the parts (1 and 3) of interest to us, nor does it, if I am not mistaken, in others.
22. I have not been able to find a proof of something like this theorem in the literature. For a possible proof, see the unabridged French version of this article (Rouilhan 2008).
23. 'This definition [...] cannot be applied to languages of a more complicated logical structure' (Tarski 1936b/1956, p. 413, n. 3).
24. In other words, $t_1$ and $t_2$ being types (in the plural), $t_1$, $t_2$ is a *single* type (in the singular).
25. It must be supposed that, for Carnap, that went without saying, since he did not say a word about it.

26. Contrary to what I did for LANGUAGE $I_0$, I shall not dub them '0-adic functors'.
27. In more common, non-Carnapian, parlance, these expressions would be predicate-expressions (predicates, respectively) *with n arguments*.
28. In more common, non-Carnapian parlance, these expressions would be *functor-expressions* (*functors*, respectively) *with n arguments*, at least when the values are of standard type.
29. Not necessarily standard, *pace* Carnap, who is not consistent on this point. Indeed, he needs the identity sign between expressions of the same non-standard type, in order, for instance, to state the axiom schema of extensionality for functions (LSL, p. 92).
30. Penned by Tarski, the last term is obviously to be taken in its ordinary, broad sense, and not in the narrow, Carnapian, sense of Carnap (1950b).
31. Curiously enough, Carnap feels duty bound to pay tribute to Gödel ('As a result of Gödel's researches it is certain, for instance, that for every arithmetical system there are *numerical properties which are not definable*' [LSL, p. 106]), even though reference to Cantor's theorem would have sufficed. Admittedly, it was Gödel who, after reading a first draft of *Syntax*, pointed out to Carnap the problem posed by the existence of undefinable properties, but it by no means justifies calling upon 'Gödel's researches'.
32. '*Numerals*' (not 'numbers'), as Carnap would have doubtlessly put it. See his definition of the concept of valuation (LSL, pp. 106–10) and what I report of it below (section 4.4).
33. Valuations are grounded here, not on natural numbers, but on numerals. Did Carnap feel more secure with the latter than with the former, or more faithful to the SYNTACTICAL character of his enterprise? Whatever the reason for the precaution might be, it seems illusory.
34. The choice of bold type, to make Kleene's definitions below easier to read, is my own. For the same reason, I also restore Carnap's use of gothic symbolism.
35. I have replaced Kleene's '$\mathfrak{S}_1 \vee \mathfrak{S}_2$ is A—***B*** (C—***B***), if $\mathfrak{S}_1$ or (and) $\mathfrak{S}_2$ is (are) A—***B*** (C—***B***)' by a clause relative to conditional '$\mathfrak{S}_1 \supset \mathfrak{S}_2$' in order to make the definition applicable to closed (and classes of closed) sentences of LANGUAGE $II_0$.
36. At the Fifth International Congress for the Unity of Science held in Cambridge, Massachusetts in 1939, Kleene distributed an abstract entitled 'On the Term "Analytic" in Logical Syntax'. He later submitted a revised and corrected version of this abstract for vol. 9 of *The Journal of Unified Science* (*Erkenntnis*). This volume was definitely prepared (Kleene's abstract figured on pp. 189–92), but, for obvious, extrinsic reasons, it was never published. I thank Paolo Mancosu and, through him, John Addison, who provided me access to this abstract. The difference between it and the corresponding part of Kleene's review of Carnap's *Syntax* is interesting, each version of Kleene's explication having its own advantages. Figuring a bit further on in the same volume of *The Journal of Symbolic Logic* in which that review appeared was a review of Kleene's abstract by Carnap (1939b). For a comparative analysis of all this material, see the unabridged French version of the present article (Rouilhan 2008). There, I devise a third explication, *à la* Kleene, having the advantages of each version he actually proposed.
37. Tarski would have said, equivalently, that it was *morphological*. Indeed, he mistakenly said this about (absolute) truth! (Tarski 1936a/1956, pp. 405–6).
38. Concerning the semantical shifting of the term, see Rouilhan (1998–9).
39. See Klein (1872) and the posthumous article of Tarski edited by J. Corcoran (Tarski 1986). In the wake of the publication of this seminal article, important work has

been done, notably by Gila Sher, Solomon Feferman, and Denis Bonnay. I still share Tarski's early scepticism.

40. See again Rouilhan (1998–9).
41. Thanks to Denis Bonnay, Serge Bozon, and Pierre Wagner for their valuable comments on an earlier version of this article. Thanks also to Claire Ortiz Hill for her help in putting my thoughts into real English.

# 5
# Carnap's Criterion of Logicality[1]

*Denis Bonnay*

Characterizing in a principled manner the distinction between logical and non-logical expressions has been a longstanding issue in the philosophy of logic. In *The Logical Syntax of Language*, Carnap proposes a syntactic solution to this problem, which aims at grounding the claim that logic and mathematics are analytic. Roughly speaking, his idea is that logic and mathematics correspond to the largest part of science for which it is possible to completely specify by 'syntactic' means which sentences are valid and which are not. Despite a renewed interest in the notion of analyticity, both inside and outside of Carnap scholarship, Carnap's definition of logical expressions has not received all the attention that it deserves. I shall argue that it is well worth a second look. More precisely, my aim will be to defend Carnap's idea against some technical problems that Carnap's original definition faces and against wider Quinean attacks on syntax-based conventionalism.

Section 1 presents Carnap's definition in the context of *The Logical Syntax of Language*, that is, how the definition exactly works and why Carnap needs it. In section 2, I review three challenges that have been raised in the literature. In section 3, I suggest a modification of the original definition. I argue that the modified version is immune to the previous challenges, and, *to some extent*, immune to new challenges as well. In the last section, I suggest that the definition has a philosophical interest of its own, because standard Quinean objections are not as conclusive as one might think when attention is paid to the fact that Carnap requires *complete* syntactic specification of validities.

## 1 Logicality in the *Logical Syntax of Language*

### 1.1 Why a definition of logical expressions is needed

In Carnap's own words, the *The Logical Syntax of Language* makes 'an attempt to provide, in the form of an exact syntactical method, the necessary tools for working out the problems of the logic of science' (LSL, p. xiii). To do so,

Carnap develops in the first three parts of the book two languages, Language I and Language II ($L_1$ and $L_2$ from now on). $L_1$ is quite restricted but $L_2$ is meant to be rich enough for classical mathematics and classical physics to be expressible in it. However, $L_2$ has no privileged status among all the languages in which this could be done. According to the principle of tolerance, there is a great variety of possible language-forms which have equal rights to be considered as the basis for the logic of science. To account for these other possible frameworks, Carnap outlines in the fourth part of the *Logical Syntax of Language* a general theory of syntax. General syntax aims at characterizing the key concepts of syntax independently of the choice of a particular language.

As part of general syntax, Carnap proposes in §50 a formal distinction between two kinds of expressions, the logical ones and the descriptive ones. For the languages $L_1$ and $L_2$, Carnap just made a list of those symbols that should count as logical.[2] But of course, no such thing can be done if one is to be concerned with the syntax of any possible language. In that case, there are just no particular expressions to list. If a distinction between logical and descriptive expressions is needed, it has to be abstractly defined in terms of what a language is, i.e. on the basis of the two sets of rules that define a language, its formation rules and its transformation rules.

Now, something like a distinction between logical and descriptive expressions is needed. In *The Logical Syntax of Language*, Carnap tries to make concrete the empiricist picture according to which the whole of science can be divided into two parts. On one side, logic and mathematics: mathematical and logical truths are *analytic*, they are not substantive and merely reflect the choice of a framework and the meaning that has been given to logical and mathematical words. On the other side, empirical science (that is, on a physicalist view, just physics broadly conceived): physical truths are *synthetic*, they are empirical truths, which can be confirmed or refuted by experience. As far as physics is concerned, the logic of science aims at making explicit these relations of confirmation or refutation. As far as logic and mathematics are concerned, the logic of science aims at establishing that there is precisely no such thing as confirmation or refutation. We are dealing here with formal auxiliaries to physics, which are deprived of any real content.

Thus Carnap has to provide a syntactic characterization of what it is to be analytic. And it should be clear that the easy answer is no answer: defining 'analytic' as 'following from the transformation rules' will just not do. Transformation rules can basically be any kind of rules. Among the transformation rules of a given language, there can be (intuitively) logical rules, like the excluded middle, but it would make perfect sense to include also as part of the framework physical rules, say, in the age of Newtonian mechanics, the three laws of motion or the law of attraction. But the law of attraction is not a formal auxiliary without any real content: far from it, this law tells us a lot about how massive bodies interact with one another. As a consequence, everything that follows from the transformation rules cannot be considered as analytic. It is perfectly okay to have physical laws as rules of transformations and as consequences thereof.[3] However, this does not turn such laws into analytic principles.[4]

Given some transformation rules, what are the *analytic* consequences of these rules? How are we to characterize those consequences of transformation rules which are fully deprived of content? Carnap has to provide an answer to this general question for his project of a general theory of syntax to succeed. And here the distinction between logical and descriptive expressions comes into play. Intuitively enough, the distinctive feature of a logical rule (from now on, L-rule, to use Carnap's jargon in the *Logical Syntax of Language*, as opposed to P-rule for 'physical rule') is that it is a rule in which no descriptive expression plays an essential role (uniform substitution of descriptive expressions preserves logical validity). An analytic consequence would then be a consequence which ensues from the L-rules alone. This is the path that Carnap follows in §51 and §52: analyticity admits of a very natural definition in terms of L-rules, and L-rules admit of a very natural definition in terms of descriptive expressions.

In the end, the burden of the definition of analyticity rests upon the distinction between logical and descriptive expressions. Knowing which expressions are logical enables one to make a difference between mere validity and genuine analyticity. In this respect, it is in the *Logical Syntax of Language* definition of logical expressions that the central claim of logical empiricism that mathematics are analytic is to be ultimately grounded. To succeed, the purported formal definition should meet two requirements:

**Descriptive adequacy:** The definition should be such that, in a given language, the (intuitively) logico-mathematical part of it does turn out to be analytic. Were it not the case, the definition would fail to show that *mathematics* are analytic, in any interesting sense of what 'mathematics' means. Conversely, (intuitively) empirical parts should turn out to be synthetic, or Carnap would be caught showing that empirical science is nothing more than a matter of framework construction.

**Explanatory adequacy:** The definition should be such that, in a given language, the analytic part of it enjoys an (intuitively) special epistemological status – like lack of content or independence from experience. Were it not the case, the definition would fail to show that mathematics are *analytic*, in any interesting sense of what 'analytic' means.

It is the well-known history of the rise and fall of logical empiricism that the project of a philosophical account of mathematics based on the notion of analyticity soon came under heavy attack: Quine's general arguments against analyticity are supposed to show that no such account is possible. It is part of a slightly less well-known history of the reception of the *Logical Syntax of Language* that the very phrasing of the definition in §50 has been shown to be problematic, so that the worm was already in the fruit at the very beginning so to speak.

Keeping in mind the two previous constraints, I shall now turn to a detailed critical examination of Carnap's formal definition of logical and descriptive expressions in §50.

## 1.2 Carnap's original definition

Here is Carnap's definition, where *S* is an arbitrary language. Gothic symbols '$\mathfrak{A}$' stand for expressions, '$\mathfrak{K}$' for classes of expressions:

> Let $\mathfrak{K}_1$ be the product of all expressional classes $\mathfrak{K}_i$ of *S* which fulfil the following four conditions:
>
> (1) If $\mathfrak{A}_1$ belongs to $\mathfrak{K}_i$, then $\mathfrak{A}_1$ is not empty and there exists a sentence which can be subdivided into partial expressions in such a way that all belong to $\mathfrak{K}_i$ and one of them is $\mathfrak{A}_1$.
>
> (2) Every sentence which can be thus subdivided into expressions of $\mathfrak{K}_i$ is determinate.[5]
>
> (3) The expressions of $\mathfrak{K}_i$ are as small as possible, that is to say, no expression belongs to $\mathfrak{K}_i$ which can be subdivided into several expressions of $\mathfrak{K}_i$.
>
> (4) $\mathfrak{K}_i$ is as comprehensive as possible, that is to say, it is not a proper subclass of a class which fulfils both (1) and (2).
>
> An *expression* is called **logical** ($\mathfrak{A}_\mathfrak{l}$) if it is capable of being subdivided into expressions of $\mathfrak{K}_1$; otherwise it is called **descriptive** ($\mathfrak{A}_\mathfrak{d}$) (1937, pp. 177–8).

The intuitive starting point is given in (2). The logical part of a language is characterized by the fact that everything is determined by the rules of transformation: there is no room for empirical confirmation or refutation, just because the transformation rules are sufficient to determine whether a sentence holds or not. As Carnap puts it:

> If we reflect that all the connections between logico-mathematical terms are independent of extra-linguistic factors, such as, for instance, empirical observations, and that they must be solely and completely determined by the transformation rules of the language, we find the formally expressible distinguishing peculiarity of logical symbols and expressions to consist in the fact that each sentence constructed solely from them is determinate. (LSL, p. 177)

A few more explanatory remarks are in order to account for the technicalities of the definition. As can be seen from (1), Carnap defines logical and descriptive as properties of expressions, not of symbols, because he considers that it is possible for a given symbol to be descriptive in certain contexts and logical in others. (4) is a natural maximality requirement: take your class as big as possible, provided determinacy is preserved. In general, there is no guarantee that there is a unique class of expressions which is the biggest one for which determinacy holds. Carnap's definition solves the problem by taking the class of logical expressions to be the intersection of maximal determinate classes.

Saunders Mac Lane, in his review of the *Logical Syntax of Language* (Mac Lane 1938), has pointed out a few minor problems, as well as ways of fixing them.[6] First, let $\mathfrak{S}_1$ be a sentence constructed out of expressions belonging to a given $\mathfrak{K}_i$. By (4), it should belong to $\mathfrak{K}_i$ as well, but by (3), it should not. However, as Mac Lane remarks, this bug could be fixed by requiring that the classes are maximal also with respect to condition (3). A somewhat similar bug threatens the modified definition. Consider a language like Language I, take two classes $\mathfrak{K}_2$ and $\mathfrak{K}_3$ such that $\mathfrak{K}_2$ is a maximal class containing all expressions of the form ∃x, but not ∃ and $\mathfrak{K}_3$ is the standard class of logical expressions. Both $\mathfrak{K}_2$ and $\mathfrak{K}_3$ satisfy requirements (1)–(4), but then neither ∃x nor ∃ can make it to the status of logical expression, because neither will be in the product $\mathfrak{K}_1$. Mac Lane suggests the following way out:

> Consider those classes $\mathfrak{K}_i$ which satisfy (1) and (2) and are maximal with respect to these conditions. For each class $\mathfrak{K}_i$, denote by $\mathfrak{L}_i$ the class of those expressions of $\mathfrak{K}_i$ which cannot be subdivided into several expressions of $\mathfrak{K}_i$, and let $\mathfrak{K}_1$ be the intersection of all $\mathfrak{L}_i$. (1938, p. 174)

From now on, I shall rely on this slightly amended version of Carnap's initial definition rather than on the initial definition itself. This fine tuning proved necessary, and Carnap can be blamed for being careless in handling the application of his definition to expressions (as opposed to symbols). But, arguably, this is no big deal, as shown by Mac Lane.

## 2 A revised definition of logicality, Carnapian in spirit

### 2.1 The misbehaviour of Carnap's definition

Unfortunately enough, the definition faces at least three more significant challenges, which shall now be considered in turn.

*Challenge 1 (Mac Lane)*

Mac Lane (1938) has an example showing that descriptive adequacy fails for any language *S* such that:

- *S* is a coordinate language.[7]
- Negation, identity as well as at least one empirical function are among the symbols of *S*.

Here is the problem. Let $f$ be an empirical function. Consider the class $\mathfrak{K}_3$ containing the expressions $f(0), =, \neg$ and all sentences constructed from these expressions (*i.e.* $f(0)=f(0), \neg f(0)=f(0)$ and so on). All these sentences are determinate. Hence $\mathfrak{K}_3$ can be extended to a maximal class. But this class cannot contain numerals, because, arguably, sentences of the form $f(0)=3$ are not determinate. As a result, numerals will not be in the product $\mathfrak{K}_1$. Thus, on Carnap's definition, numerals will not come out as logical expressions, even for a language like Language II.

*Challenge 2 (Creath)*

Creath (1996) has another counterexample to descriptive adequacy. The problem arises in any language *S* such that:

- *S* contains an empirical predicate and an ostensive definition for that predicate.

Consider an empirical predicate *P* equipped with an ostensive definition '*Pa*' sanctioned by the appropriate transformation rules. '*P*' could be 'has a weight of one kilogram' and '*a*', 'the international prototype kilogram'. '*Pa*' is an ostensive definition because, assuming that we know what it is for two objects to have the same weight, pointing at the prototype as satisfying the property of *P* completely fixes what we mean by 'one kilogram'. Thus it does make sense to have '*Pa*' among the transformation rules of our language.

So by hypothesis, '*Pa*' is valid for some object, or for that matter, position, *a*. Now *Pa* alone is determinate, so there is a maximal determinate class $\mathfrak{K}_4$ containing *P* and *a*. But $\mathfrak{K}_4$ is bound to be quite small. Arguably, existential quantification and identity cannot be in $\mathfrak{K}_4$: there is no reason why sentences like 'there are *n* objects weighing one kilogram' should be determinate. Similarly, if *S* is a coordinate-language, numerals cannot be in $\mathfrak{K}_4$: there is no reason why sentences like '*Pb*' (where *b* is a given position) should be determinate. Again, by construction, expressions excluded from $\mathfrak{K}_4$ will be eventually excluded from $\mathfrak{K}_1$.[8]

*Challenge 3 (Quine)*

Notwithstanding some more far-reaching objections, Quine (1963) makes a point against Carnap's definition, by highlighting that descriptive adequacy fails for any language *S* such that:

- *S* is not a coordinate-language.
- *S* contains an empirical predicate, and some very general properties of this predicate are encapsulated in the transformation rules for *S*.

Quine considers a predicate *HR* standing for 'heavier than'. Assuming that numerals are not used as position names, *HR* could be added to the logical part of *S* while preserving determinacy: it would be sufficient to enrich a bit the transformation rules so that sentences like '$\exists x, y\ HRxy$' come out as valid. As a result *HR* is bound to qualify as a logical expression, which is of course an unwelcome consequence! Quine remarks that in a coordinate-language, where position names are logical expressions, the problem will not arise because *HRab* is not in general determinate for arbitrary coordinates *a* and *b*.

These three cases against the adequacy of Carnap's definition are quite compelling and even more so when they are considered together. Challenge 3 suggests that a coordinate-language is necessary for the definition to work. But challenge 1 establishes that numerals will not in general come out as logical in a coordinate-language. And, by making the slightly stronger assumption that *S* has an ostensive definition for an empirical predicate, challenge 2 raises a problem which does not even depend on coordinates being used or not.

## 2.2 Managing misbehaviour

Is it possible to draw a positive lesson from these challenges that would suggest how to patch the definition? Note that challenges 1 and 2 share a common structure. In both cases, descriptive inadequacy ensues from the dramatic shrinking of logical expressions which results from taking the *intersection* of maximal classes. What happens is that an empirical predicate sneaks in in one of the $\mathfrak{K}_i$, and prevents that $\mathfrak{K}_i$ from being extended to a reasonable maximal class. Challenge 3 is similar in this respect, though it exhibits a misclassification of an empirical predicate as logical rather than the other way around. As a matter of fact, challenge 3 shows how an empirical predicate can creep in and belong to the biggest determinate class in the absence of coordinates.

This suggests the following diagnosis. Taking the intersection is necessary to gain uniqueness. Undue appearances of empirical predicates in the $\mathfrak{K}_i$ make the price of uniqueness unbearably high. Still, it would be a shame to renounce uniqueness: the idea that there could be several distinct types of logical expressions in one and the same language seems just too weird. An expression would be logical (compatible with determinacy) in one context, but not in another. But, if that expression was genuinely logical, it should be compatible with determinacy in all contexts, that is, it should always be possible to add it to a set of expressions yielding a determinate class of sentences and to preserve determinacy of the corresponding class. Moreover, challenge 3 makes a point which is independent of the uniqueness requirement.

So the problem is definitely with empirical predicates sneaking in and making their way into the $\mathfrak{K}_i$. The empirical function of Mac Lane, the ostensive predicate of Creath and Quine's 'heavier than' should not be allowed membership in a class from which $\mathfrak{K}_i$ is to be constructed. But why does this happen? The application of these expressions, which, by hypothesis, are empirical, is indeed indeterminate. Therefore, intuitively, they are not supposed to appear in a determinate class. But they do. This is because the guilty classes are in some sense arbitrarily restricted. They do not encompass enough position or object names for the indeterminacy to shine through. A solution suggests itself: testing for determinacy should always occur in a context in which there are enough names for the test to be significant. Building on this intuition, I propose to modify the original definition along the following lines:

Let $\mathfrak{N}$ be the class of names. Consider those classes $\mathfrak{K}_i$ such that:

(1) If $\mathfrak{A}_1$ belongs to $\mathfrak{K}_i$, then $\mathfrak{A}_1$ is not empty and there exists a sentence which can be subdivided into partial expressions in such a way that all belong to $\mathfrak{K}_i \cup \mathfrak{N}$ and one of them is $\mathfrak{A}_1$.
(2) Every sentence which can be thus subdivided into expressions of $\mathfrak{K}_i \cup \mathfrak{N}$ is determinate.
(3) $\mathfrak{K}_i \cap \mathfrak{N} = \varnothing$

and which are maximal with respect to (1) and (2). For each class $\mathfrak{K}_i$, denote by $\mathfrak{L}_i$ the class of those expressions of $\mathfrak{K}_i$ which cannot be subdivided into several expressions of $\mathfrak{K}_i$, and let $\mathfrak{K}_1$ be the intersection of all $\mathfrak{L}_i$.

This modified definition meets the three previous challenges. Concerning challenge 1, the empirical function $f$ will not belong to any of the $\mathfrak{K}_i$. As an empirical function, $f(a)=b$ will be indeterminate for some $a, b \in \mathfrak{N}$. The same thing happens for challenge 2. Even though the empirical predicate $P$ comes with an ostensive definition which says that '$Pa$' holds for some object $a$, there will still be some $b \in \mathfrak{N}$ such that $Pb$ is not determinate. Similarly regarding challenge 3, 'heavier than' will be indeterminate for some $a, b \in \mathfrak{N}$.

What about the members of $\mathfrak{N}$ themselves? The previous definition does not say anything about them. Maybe this is as it should be. In a coordinate-language, it seems that deciding that numerals are logical or descriptive symbols does not make much sense. A numeral can be used as a name for a number, as in '$18+17=35$': in that kind of context, it is logical rather than descriptive (intuitively). But in a coordinate-language, it can also be used as a name for a position, as in '*Red*(18)' ('the position 18 is red'): in that kind of context, it is descriptive rather than logical (intuitively). But this dispute need not be addressed. As recalled in the first section, the distinction is here to make a definition of analyticity possible. And we should be happy if it is possible to do so without answering the question concerning the status of members of $\mathfrak{N}$. Now, the thing is that this leaves two options for the definition of L-validity (or analyticity):

- either a sentence $\mathfrak{S}$ is **L-valid** iff it is valid and so is every other sentence $\mathfrak{S}'$ obtained from $\mathfrak{S}$ by uniform replacement of every symbol which is not in $\mathfrak{K}_1$ by an expression of the same genus,
- or a sentence $\mathfrak{S}$ is **L-valid** iff it is valid and so is every other sentence $\mathfrak{S}'$ obtained from $\mathfrak{S}$ by uniform replacement of every symbol which is not in $\mathfrak{K}_1 \cup \mathfrak{N}$ by an expression of the same genus.

The first option will not work. In $L_2$ for example, '$2+2=4$' is valid, but '$3+3=4$' is not, hence '$2+2=4$' would not qualify as an L-validity. Of course, on the second definition and under the assumption that '+' and '=' turn out to be logical – as they should – '$2+2=4$' will be L-valid – as it should. Note that there is no dual problem with the second attempt. Let *S* be a language such that '*Pa*' is valid, say as an ostensive definition. '*Pa*' will not be L-valid, because there will clearly be a sentential function $\phi(x)$ such that $\phi(a)$ is not valid. Thus, the good notion of validity associated with our revised definition of logical expressions seems to be given by the second definition. In a sense, this amounts to implicitly treating names as logical symbols. But this should be not considered as completely *ad hoc*, since elements of $\mathfrak{N}$ are, by construction, compatible with determinacy.

What has been suggested so far is that there is a common source to the failures of descriptive adequacy presented by the three challenges to Carnap's definition: empirical predicates can sneak in and make their way to maximal determinate classes because the determinacy test does not involve a sufficiently wide range of names. Accordingly, a modification of the definition has been suggested, which forces predicates to be tested for determinacy against all names in the language. This modified definition does meet the three challenges. Is it everything we can ask for? One might object that Mac Lane, Creath, and Quine's criticisms suggest that there is definitely something wrong with the kind of syntactic attempt at a definition of logical expressions in which Carnap engaged. I have by no means proved that testing against all names will block any kind of counterexample. And one might think that chances are high that it will be possible to devise some other kind of clever counterexample to the descriptive adequacy of the modified definition.

Well, first, it is just not clear what a proof of the descriptive adequacy of the new definition would look like. But it seems fair to say that producing such a definition reverses the burden of the proof. Facing the three previous challenges, someone who is sympathetic with Carnap's intuition had to show that the definition could be amended. In so far as the previous counterexamples do not apply to the amended definition, someone who thinks that any Carnapian attempt based on determinacy is bound to fail has to offer a challenge analogous to those that have been discussed.

Nevertheless, one could ask whether the new definition works for the kind of languages that Carnap considers. After all, a striking feature of the challenges we discussed is that they did not resort to exotic languages. On the contrary, they took place in the very framework of languages like $L_1$ and $L_2$ with added empirical predicates.

Let $S$ be a first order coordinate-language based on the following vocabulary:

- a set $\mathfrak{N}$ of numerals,
- a set $\mathfrak{M}$ of symbols for first-order logic and arithmetic (those needed on top of the numerals),
- a set $\mathfrak{E}$ of (first-order) empirical predicates.[9]

The transformation rules of $S$ will typically include logical rules, rules for first-order arithmetic plus rules corresponding to a physical theory for the $\mathfrak{E}$ symbols using $\mathfrak{M}$ – these rules may include general laws as well as the kind of ostensive definitions Creath introduces. Note that this language is richer than $L_1$, but poorer than $L_2$ (which is a higher-order language). The following fact holds:[10]

**Fact:** If the transformation rules of $S$ are such that:

(a) the $\omega$-rule is admissible for the whole language, and the arithmetical rules are strong enough to guarantee $\Delta_0$-completeness
(b) for every $n$-ary predicate $P$ in $\mathfrak{E}$ there are numerals $a_1, \ldots, a_n$ in $\mathfrak{N}$ such that $Pa_1 \ldots a_n$ is not determinate

then all symbols in $\mathfrak{M}$ are logical and all symbols in $\mathfrak{E}$ are descriptive.

Condition (a) guarantees that the transformation rules are strong enough so that what should be determinate is determinate. If the $\omega$-rule is not admissible, it can be the case that all substitution instances $\phi(0), \ldots \phi(n), \ldots$ of a formula $\phi(x)$ are valid though $\forall x\ \phi(x)$ is not valid. But if all the instances of $\phi(x)$ are determinate, we clearly expect $\forall x\ \phi(x)$ to be determinate as well. If it were not the case, a demarcation of logical expressions based on determinacy might yield unwelcome results just because the property of determinacy itself is badly implemented in the language.[11] Condition (b) guarantees that the transformation rules are weak enough so that what should not be determinate is not determinate. If condition (b) did not hold, the applicability of some empirical predicate to all numerals would be determinate: but this should not happen to an empirical predicate. To sum up, if either condition (a) or (b) fails, we cannot expect a definition of logical expressions based on determinacy to work, because the very connection between determinacy and logicality is undermined in the definition of the transformation rules. Conversely, it seems that any good definition of logical expressions should work when conditions (a) and (b) are satisfied. The previous Fact says that this is indeed the case for the modified definition: logical and mathematical symbols are classified as logical expressions and other predicates are classified as descriptive.

Thus, it is possible to show that the suggested modification of Carnap's definition does succeed in handling not only the three specific challenges considered but also all similar challenges that could be constructed in a first-order setting. Note, however, that the restriction to a first-order language is substantial. In a language with higher-order empirical predicates, our three challenges could be reproduced at a higher level, because testing on all names would be useless against counterexamples involving such predicates.[12] In order to deal with such cases, the definition would have to be modified in order to extend 'systematic tests' to higher-order levels.[13]

Therefore, the modified definition should not be taken as a final victory: it has not been established that it is possible to provide a syntactic definition matching the level of generality required by the endeavour of a theory of general syntax. But significant progress has been made. The previous challenges showed that Carnap's definition did not work even for fairly simple languages – mild extensions of Language I with empirical predicates. Now, the previous Fact shows that the new definition does work for non-trivial languages containing arithmetic plus empirical predicates and transformation rules for those predicates.

## 3 Analyticity and determinacy

### 3.1 Objections against syntactic definitions of analyticity

So far so good. It is possible to make good (technical) sense of Carnap's definition of analyticity in terms of determinacy so that, up to some point, descriptive adequacy is secured. But what about the explanatory value of Carnap's definition? Does the determinacy criterion account for the fact that L-validities enjoy a special epistemological status – they are a matter of choice of framework, rather than a matter of what is true in a given framework? Of course, the answer depends on what is at stake. What does it mean to say that logical and mathematical truths are 'true by convention'? Or, what exactly is Carnap's aim in showing that mathematics are analytic? Much ink has been spilled on these issues, and I shall engage myself neither in a full-blown defence of conventionalism nor in a reconstruction of Carnap's position.[14] Rather, I shall set modest success conditions for Carnap's definition. According to the criticisms that Quine directs at the *Logical Syntax of Language* in his paper 'Carnap and Logical Truth', Carnap is bound to fail, no matter how modest the success conditions are. Quine's criticisms should be valid no matter what is precisely Carnap's aim, because his point is that Carnap simply fails to draw any distinction, so that 'no special trait of logic and mathematics has been singled out after all' (Quine 1976, p. 125). Carnap's definition would (modestly) succeed if it were to bring up such a special trait, even if the philosophical significance of that trait for his overall project might in turn be debatable. And I want to argue that determinacy does single out a special trait of logic and mathematics.

According to Quine, requiring that logico-mathematical truths are specifiable in syntactical terms does not make for a distinction between logico-mathematical truths and other kinds of truths. His criticism is twofold. First, Quine has a concern with what syntax means in this context. For languages like $L_2$, which contain a rich logico-mathematical theory, Carnap has to use transformation rules, such as the $\omega$-rule, which are far from elementary to guarantee that logic and mathematics are analytic. In general, in order to make logic and mathematics syntactically determinate, it is required to use a metatheory for the definition of the syntax which uses all of logic and mathematics (actually, stronger mathematics are needed in the metalanguage). But what is the point of showing that logico-mathematical truths can be specified in logico-mathematical terms? This is far from being a *tour de force*: the same thing can be done for any other theory. For example, it is not a problem to lay down the sentences which are recognized as true by our best physical theory using physics in the metatheory.

Then Quine has a concern with specifiability itself, be it by elementary syntactical means or not. After all, any finite set of truths can be specified by syntactical means, even by narrowly syntactical means. Does this make these truths conventional in any interesting sense? Imagine that an arbitrary part of our physical theory of the world is reproduced by syntactical means. Does this make this part of physics any less empirical and any more conventional?

To sum up, Quine doubts the significance of Carnap's formal reconstructions when it comes to showing that logic and mathematics enjoy a special epistemological status. According to Quine, one should realize that, in this respect, nothing can be done for logic and mathematics which cannot be done for physics or economics. Strangely enough though, Quine's criticisms focus on the possibility of reproducing by formal means a given class of truths; and it seems that he does not take into account the supplementary requirement introduced by Carnap, namely determinacy.

## 3.2 The significance of determinacy

Let us grant that there is nothing special with being specifiable by means of transformation rules. As Quine puts it, physical truths according to a given physical theory are specifiable in a notation consisting solely of names of signs, operators expressing concatenations of expressions, *and* the whole physical (and logico-mathematical) vocabulary itself. This is because 'Tarski's routine of truth definition [will] still carry through just as in the case of logic and mathematics' (Quine 1976, p. 125). But Quine overlooks a major difference here. Using a strong metatheory, Carnap can show a determinacy result for the sublanguage of logic and mathematics in $L_1$ and $L_2$. Nothing like this is to be expected in the case of physics: there will be cases in which the question whether, say, there is a such and such particle at such and such position in space-time, is left undecided by the theory. And this has nothing to do with the need for a stronger metatheory, as in the case of logic and mathematics. Physics is about finding some general laws and applying them to some particular situations in order to be able to predict how these situations evolve; but physics does not provide us with a complete picture of physical truths in the same way as a complete axiomatization of number theory provides us with a complete picture of arithmetical truth. These seem to be facts about what mathematics and physics are, as we practise them.

Do Quinean objections still apply? Maybe one could say that physics strives for a complete theory, so that there is actually no *principled* difference between physics and mathematics. After all, this is the very idea of Laplace's demon: if determinism is true,[15] knowledge of all forces that set nature in motion plus a complete description of all positions of all items of which nature is composed would result in complete knowledge of what has been or will be. As a result, if the transformation rules for a language *S* are set up by Laplace's demon, physics in *S* will be as determinate as mathematics in *S*, hence, on Carnap's criterion, as analytic as mathematics. Again, this would show that no special trait of mathematics has been singled out after all. However, it is less clear that Carnap has a situation here. All the truths of a given physical theory *are* specifiable by syntactic means, broadly conceived, so requiring specifiability is not sufficient. But, given that physics *could be* extended to a complete theory, does it follow that determinacy is not enough? From the point of view of Laplace's demon, mathematics and physics are about the same: the chances for surprise are zero in both fields. By hypothesis, physics for the demon can now go without experience. So it might not be an unwelcome consequence of Carnap's approach that it classifies such a physics as analytic. Physics is exhaustively hardwired in the framework, and in that respect, the framework does make experience irrelevant to physics. The collapse of mathematics and physics in the counterfactual situation in which transformation rules are laid down by the demon is not a problem for Carnap, because, in such a situation, it seems intuitive enough to consider that the distinction between mathematics and physics, as far as their relationships with experience is considered, has indeed disappeared.

A different objection may be derived from Quine's worries about Carnap's very liberal use of the notion of syntax. Now, imagine that I fancy pretending I am Laplace's demon. I define the transformation rules for a language *S* by setting that $\phi$ is an axiom (in the terms of the *Logical Syntax of Language*, 'a primitive sentence') if and only if $\phi$ is true. By the law of the excluded middle, every sentence $\phi$ is true or false, hence, trivially, every sentence in $S$ is determinate. Obviously, this is cheating. I am not Laplace's demon, I am just pretending. As a matter of fact, there are plenty of questions about observational and theoretical properties that I would be unable to answer. So I did not really specify which sentences $\phi$ I am taking as axioms. But what does it mean to provide a *real* specification? One could remark that I gave no mechanical procedure which would generate the set of sentences that are valid in $S$. That's true. But, so goes the objection, neither did Carnap for his language $L_2$. By Gödel's theorem, the set of arithmetical truths is not recursively enumerable, so it is not possible both to secure determinacy for arithmetic and to provide a mechanical procedure generating the set of valid sentences. So Carnap would be cheating, just as I did when I was pretending to be as smart and knowledgeable as Laplace's demon. To put it another way, Carnap bypasses Gödel's theorem by resorting to transfinite methods (the $\omega$-rule and worse). But this is specification by *fiat*, and specification by *fiat*, uninteresting as it is, could be used to specify the set of physical truths as well.

However, the objection rests on the debatable assumption that validities have to be recursively enumerable. In a foundational approach, that is, if the aim was to reduce mathematical truths to some more elementary truths, the requirement would be quite sensible. But Carnap's approach is clearly not foundational in that sense. Therefore, the fact that a stronger metatheory is needed to define logico-mathematical validities does not seem to give a knock-down argument. There is a price to pay for determinacy, but it is not clear that the price is too high for Carnap. The extra power which is needed to specify arithmetical truths is mathematically tractable (for example, using generalized recursion theory) and it seems just not true to assimilate that kind of specification to saying something like '$\phi$ is an axiom if and only if $\phi$ is true'. Moreover, note that this problem only concerns complex arithmetical sentences: the class of atomic arithmetic sentences is decidable, which certainly makes a big difference with the class of atomic sentences of physics. So there would be another way to deal with the previous criticism by putting a stress on determinacy for simple sentences.

It seems thus that Carnap's criterion, or the modified version we proposed, can resist Quine's criticisms. Indeed, this is what Carnap himself thought, judging from his shorthand reading notes on Quine's *Truth by Convention*, as reported by Creath (1987):

> One can conventionally lay down geometry and physics, and thereby stipulate or restrict the meaning of the empirical primitive terms.

> Problem: In what specific sense are logic and mathematics then «conventional», in contradistinction to physics? He [Quine] means there is no clear boundary. Therefore it is difficult to see what the thesis of the conventional character of logic and mathematics (contrary to other areas) amounts to. To this I [say]: it seems to me that the difference lies in that the stipulations in logic and mathematics can be so effected that the truth of all true logical and mathematical sentences is laid down. [ . . . ] Against this, we can never lay down physics so that the truth of all true sentences is laid down. (ASP, RC 102-61-06; transcribed from German shorthand by R. Nollan and translated from German by R. Creath)

Commenting on this quote, Creath remarks that, despite the plausibility of this strategy, Carnap's attempt at spelling it out fails for technical reasons. I agree with Creath that Carnap's strategy should not be underestimated. And when it comes to the technical difficulties it faces, the situation is less desperate than it might seem.

## 4 Conclusion

To conclude, Carnap's determinacy criterion for logicality surely deserves more attention. For various reasons, it seems that the impact of determinacy in a characterization of the distinction between logical and descriptive expressions has been overlooked. Among these reasons, one might mention Carnap's shift toward semantics, Quine's general arguments against analyticity, as well as the technical problems of the definition in the *Logical Syntax of Language*. I have tried to show that these technical problems can be overcome. There surely is a risk of revenge. But, the burden of proof is now on Carnap's opponents. Similarly concerning the philosophical import of determinacy, standard Quinean arguments against conventionalism do not go through. This is no vindication of conventionalism. But, again, interesting prospects seem to be open. The syntactic route faces other problems, and it yields particularly awkward definitions for the transformation rules when higher-order languages like $L_2$ come into play. Therefore, it would be particularly interesting to see whether determinacy can be made sense of in a semantic framework, and how Carnap's criterion compares with semantic approaches to logicality.[16]

### Appendix

We recall here the Fact to prove, which says that the revised definitions of logical expressions work for (at least) some category of languages:

**Fact:**

Let $S$ be a first-order coordinate-language, whose vocabulary is based on a set $\mathfrak{N}$ of numerals, a set $\mathfrak{M}$ of additional symbols for first-order logic and arithmetic and a set $\mathfrak{E}$ of (first-order) empirical predicates, if the transformation rules of $S$ are such that:

(a) the $\omega$-rule is admissible for the whole language, and the arithmetical rules are strong enough to guarantee $\Delta_0$-completeness
(b) for every $n$-ary predicate $P$ in $\mathfrak{E}$ there are numerals $a_1,\ldots,a_n$ in $\mathfrak{N}$ such that $Pa_1\ldots a_n$ is not determinate

then all symbols in $\mathfrak{M}$ are logical and all symbols in $\mathfrak{E}$ are descriptive.

**Proof:**

(a) All symbols in $\mathfrak{M}$ are logical.

Let $\mathfrak{K}'_i$ be a class satisfying constraints (1) and (2) of the (new) definition. It is sufficient to show that any maximal class $\mathfrak{K}_i$ extending $\mathfrak{K}'_i$ is such that $\mathfrak{M} \subseteq \mathfrak{K}_i$. Let us say that a sentence $\phi$ is atomic′ iff it is built out of $\mathfrak{K}'_i \cup \mathfrak{N}$ and such that there is no subexpression of $\phi$ which is a sentence that can be built out of $\mathfrak{K}_i \cup \mathfrak{N}$. By condition (2), every atomic′ sentence is determinate. By hypothesis (a), we have $\Delta_0$-completeness for arithmetical sentences, so determinacy of atomic′ sentences will be preserved if we add arithmetical predicates. Now, again by hypothesis (a), the $\omega$-rule is admissible for the whole language, so, by induction, determinacy for atomic′ sentences extends to all sentences that can be built by logical means out of the atomic′ sentences. Hence every sentence that can be subdivided into expressions of $\mathfrak{K}_i \cup \mathfrak{N} \cup \mathfrak{M}$ is determinate. By the maximality of $\mathfrak{K}'_i$, this implies that $\mathfrak{M} \subseteq \mathfrak{K}_i$.

(b) All symbols in $\mathfrak{E}$ are descriptive.

This is a straightforward consequence of hypothesis (b). Let $P$ be an $n$-ary predicate in $\mathfrak{E}$. Assume $P$ is logical. $P$ belongs to the maximal classes $\mathfrak{K}_i$. But then by definition of those classes $\mathfrak{K}_i$, every sentence based on $\{P\} \cup \mathfrak{N}$ has to be determined. By hypothesis (b), there are numerals $a_1,\ldots a_n$ in $\mathfrak{N}$ such that $Pa_1\ldots a_n$ is not determinate. Contradiction. Therefore $P$ is descriptive. QED.

## Notes

1. This chapter originates from discussions held during a reading seminar on Carnap's *Logical Syntax of Language*, which was organized at the IHPST by Pierre Wagner. I would like to thank the participants, and especially Serge Bozon, Philippe de Rouilhan, and Pierre Wagner for their helpful comments. A preliminary version was presented during a conference at the IHPST in 2005. I would like to thank the audience for its accurate remarks, and particularly Richard Creath, whose 1996 paper convinced me to take a fresh look at Carnap's definition of logical expressions, and Steve Awodey, for some stimulating criticisms and for drawing my attention to Saunders Mac Lane's review of Carnap's book.
2. More precisely, for $L_1$ as example, 'the logical primitive symbols [...] consist of the eleven individual symbols mentioned already, together with nu and all the $\mathfrak{z}$'

(p. 23), where the eleven individual symbols consist of punctuation marks (parentheses, commas) and the 'standard' logical symbols (propositional connectives and quantifiers), 𝔫𝔲 is the numeral for zero and the 𝔷 are numerical variables. In contrast, primitive predicates and functors are descriptive symbols, and an explicitly defined symbol is descriptive if a primitive predicate or functor appears in its definition-chain (p. 25). The inclusion of punctuation marks among logical symbols might look weird from a contemporary perspective; these are rather considered as 'auxiliary symbols' for which the distinction logical *vs.* descriptive does not make sense. One reason for thinking so is that auxiliary symbols are deprived of semantic content; but in the purely syntactic setting of *The Logical Syntax of Language*, this point misses the mark. Moreover, classifying auxiliary symbols as logical is consistent with the general definition to be provided in the general syntax. On Carnap's use of gothic letters, see above, Introduction, pp. 43–5.

3. Actually, as Carnap himself remarks, 'it is possible to go even further and include not only universal but also concrete sentences – such as empirical observation sentences. In the most extreme case, we may even so extend the transformation rules of S that every sentence which is momentarily acknowledged [...] is valid in S' (LSL, p. 180).
4. One could ask whether the conceptual possibility of putting some physical laws among the transformation rules is worth the fuss. Creath (1996, p. 254) argues that Carnap is interested in P-rules as 'a way to reconcile the two major wings of the Vienna Circle on the issue of the nature of theories'. Note, however, that the mere conceptual possibility does matter: it shows that validity cannot be all there is to analyticity, because treating a given physical law as a P-rule certainly does not make it deprived of content or true by mere convention. More on this below.
5. According to Carnap, a sentence 𝔖 is determinate in S iff either it is valid in S or every sentence is a consequence of it, that is to say, 𝔖 is contravalid.
6. He also points out what he takes to be a rather major problem. This problem will be considered separately in the next section.
7. Carnap uses the label 'coordinate-language' for languages which designate objects by systematic positional coordinates rather than by proper names. In such a language, number names are used as names for positions as well as names for numbers, so to speak. As a consequence, empirical predicates are applied to number names. For example, 'temp(17) = 15' will mean that the object located at position 17 has temperature 15.
8. Our presentation of the argument differs from Creath's. Creath considers that 'the initial problem that [he is] describing does not arise for Carnap's Language I and Language II in the *Logical Syntax of Language*. This is because these are position languages' (1996, p. 265). Creath notes that in a position language in which 2 is constructed as $0''$, the position name involved in an ostensive definition will not be a symbol but rather a complex expression. However, Carnap's definition does apply at the level of expressions, so that, in any case, the maximal class extending '*Pa*' will be a legitimate class to consider.
9. For the sake of simplicity, we do not allow for empirical functions. Of course, this is no loss of generality, because they can be replaced by empirical predicates. The character '𝔈' is to be read 'E'.
10. See proof in the Appendix.
11. Actually, this is precisely the reason why Carnap had to use a version of the $\omega$-rule in his languages.

12. On the contrary, the restriction to a coordinate-language could be suppressed, provided that the requirement of admissibility of something like an $\omega$-rule is 'extended' to cover all names in the language.
13. There should not be any major difficulty here, though the definition would certainly become much more complicated.
14. Recently, at least two ways of rescuing Carnap's programme in the *Logical Syntax of Language* from Quine's objections have been explored. The first strategy consists in proposing a weak reading of Carnap's programme, according to which it could succeed even if determining which sentences are analytic is arbitrary (see for example O'Grady [1999]; O'Grady so characterizes Carnap's position that it escapes Quine's criticisms, p. 1015). In contrast, the second strategy aims at refuting directly Quine's arguments to the effect that determining which sentences are analytic is arbitrary (see Chapuis-Schmitz [2006], who articulates such a proposal and offers a survey of the literature).
15. For the sake of the argument, I will suppose that determinism is a reasonable assumption, but note that the present objection falls if determinism is false.
16. Following a remark by Steve Awodey, invariance under permutation, which has been promoted by Tarski as a criterion for logicality in the semantic setting, results in purely logical sentences having the same truth-value in all models of the same size. This suggests an interesting convergence between Tarski's and Carnap's approaches to logicality.

# Part III

# Carnap's Philosophical Programme and Traditional Philosophy

# 6
# Logical Syntax, Quasi-Syntax, and Philosophy

*Jacques Bouveresse*

> If I had to point out what I regard as the greatest single achievement of Logical Empiricism (and of Analytical Philosophy in general), I would not hesitate to declare that this greatest achievement consists in establishing and corroborating the thesis that many, if not most, philosophical controversies are not, as they are commonly regarded by participants and onlookers alike, theoretical disagreements on questions of fact (of a scientific, or ethical, or aesthetical, or ... nature) but rather disagreements [...] on the kind of linguistic framework to be preferably used in a certain context and for certain purpose. (Bar-Hillel 1963, p. 533). Cf. Carnap (1963a, pp. 941–2)

## 1 The principle of tolerance and the rejection of the 'factuality of meaning'

In his book, *The Semantic Tradition from Kant to Carnap* (1991), Alberto Coffa suggests that the Carnapian principle of tolerance, as it is defended in *Logical Syntax of Language*, has a major weakness. The weakness is constituted by the link the principle has with the complete rejection of what can be called 'the factuality of meaning'. For the representatives of that position, there can be nothing factual about meaning, there can be meaning only by convention, and then there can be truth in virtue of meaning, but there cannot be truth about meaning and about the conditions of meaningfulness and meaninglessness in general. 'The worst side of the principle [of tolerance] embodies', says Coffa, 'the semantic conventionalism that we have just encountered in Reichenbach and Popper, the idea that in matters of meaning there is nothing interesting to discover and everything to decide upon' (Coffa 1991, p. 320).

Since, according to Carnap, it is entirely a matter of convention to choose to use a language whose rules of inference are purely logical or also to include 'physical' rules, it is not even correct to affirm, as Wittgenstein does in the *Tractatus*, that 'it is one of the most important facts that the truth or falsehood of non-logical sentences cannot be recognized from the sentence alone' (1922, 6.113). To say that is, for Carnap, to defend an unacceptable form of 'absolutism'. For, that too is not and cannot be a fact: if we have decided to add some P-rules to the L-rules of our language, a question which is not itself one of truth or falsehood, but only of convention, it may very well happen that the truth or falsity of certain synthetic sentences can also be recognized from the form alone (*The Logical Syntax of Language*, p. 186. LSL henceforward). Wittgenstein, according to Carnap, 'leaves out the conventional factor in language-construction' (LSL, ibid.). We may, of course, to say the least, wonder whether that is really what Wittgenstein does, but what is quite clear is that the only factor Carnap himself seems to be willing to take into account is the one which he reproaches Wittgenstein for having ignored, i.e. the conventional one.

But this is, obviously, not the whole of the story. One of the main problems Carnap had, according to Coffa, with the principle of tolerance was that some of the convictions he wanted to express in his book, especially those concerning philosophy, had a clearly non-conventional character and that he was led inevitably to formulate them, if not overtly, at least inadvertently, in a non-conventional way. Coffa maintains that 'there is no coherent reading of LSL that takes seriously the semantic conventionalism in that book' (1991, p. 322). When Carnap says, for example, that the problems of foundations and other philosophical problems are 'at bottom [*im Grund*] syntactical, although the ordinary formulation of the problems often disguises their character' (LSL, p. 311), he implicitly asserts, in a way that cannot be other than factual, that the philosophical questions and the philosophical statements have a true character and a proper formulation, which can only make them appear as syntactical sentences concerning a language already specified or still to be specified. He evidently does not content himself with saying that he has adopted, and proposed to all the people who could be willing to follow him to adopt, a convention according to which the propositions of the philosophers have to be considered and reconstructed in that way. As Coffa says:

> it is impossible to take seriously the view that the thesis of metalogic, the doctrine that all philosophy is about language, is no more than a proposal, an invitation to look at things from a certain perspective; or to believe that Carnap's painstaking constructions of languages for constructivists and classicists were not really inspired at least by the suspicion that other philosophers had in fact misunderstood the situation. (1991, p. 322)

It is, therefore, difficult, not to say impossible, to give a coherent interpretation of Carnap's position without attributing to him what could be called a 'second-level semantic factualism', which is, at least, an implicit doctrine of LSL. In the case of the philosophy of mathematics, for example, there must clearly be, behind the first-level semantic conventionalism, a second level of semantic factualism that constitutes a real challenge to all the other philosophies of mathematics and is, in fact, never explicitly endorsed by Carnap. According to Coffa:

> Carnap was proposing first an object-language conventionalism, arguing that, if you accept and I reject, say, the multiplicative axiom, we are not disagreeing on a matter of fact, however ethereal the fact, but are following different paths in the characterization of the language we intend to use. My acceptance of that axiom is not a manifestation of the fact that I have identified a true statement, but part of the process through which I identify the language I will use. The multiplicative axiom is not a factual claim, but a convention. But *this* statement is not a proposal for a convention. It is a factual claim about the nature of mathematical axioms. This is the second-level factualism, the presupposition that there is a fact of the matter concerning the difference between the stage at which we produce the semantic machinery involved in communication and the stage at which we are finally communicating – or, if you will, the analytic-synthetic distinction. The role of sentences in the former stage is, as we know, the key to the new theory of the a priori. (ibid.)

The 'thesis of metalogic', as it could be called, says that philosophical propositions are not about the world, but about the structure of an object-language in which we talk about the world. It was first stated in Carnap (1932b) in the following terms: 'The meaningful philosophical propositions are metalogical propositions, i.e. they deal with forms of language' (p. 435). If the formal mode of speech has to be preferred to the material one, that is because the latter gives a completely misleading idea of what propositions of that kind are and what they say. But Coffa is certainly right when he remarks that, if the material mode is misleading, that can only be because it suggests something that is *false*, and if the formal mode is better, that can only be because it tells us the *truth* about the real situation. The thesis according to which the philosophical propositions deal not with the world, and not even with *the* language (for to say that would be to adopt a dogmatic and absolutistic point of view), but with forms of language or linguistic frameworks of different kinds, tells us something about what they really mean and it invites us, it seems, to recognize a fact of some kind, and not simply to give our agreement to a proposed convention.

Carnap calls 'pseudo-object sentences' and also 'quasi-syntactical sentences of the material mode of speech' those of the philosophical propositions which are not the product of logico-grammatical confusion, i.e. they are not simply nonsense, but really tell us something. Only, they do not discuss what they mistakenly give the impression of speaking about, but something different. In the language of the *Logical Syntax of Language*, we can say that they are formulated in a *transposed* –and fallaciously transposed – mode of speech. A philosopher is, of course, free to accept or to reject an assertion of that kind and, as many philosophers would certainly do, can defend a completely different conception of what philosophical propositions are. Carnap was, of course, quite conscious of that, and particularly anxious, in philosophy, to avoid giving any impression of dogmatic self-confidence, even on the crucial question of the real status of philosophical propositions, but that does not mean that he would have been ready to concede that we are free to accept or to reject a proposition like the thesis of metalogic exactly in the same way as we are free to accept or to reject a convention, even if it would certainly be difficult, not to say impossible, to prove that it corresponds more closely than the usual conception to the true 'nature' of philosophy.

Coffa observes that: 'Carnap's attitude toward philosophical considerations was roughly that of the scalded cat toward boiling water. He was second to none in his ability to state clearly and argue cogently formal-level philosophical issues; but the deeper and less obviously formal those issues become, the harder it is to find either a clear statement or an argument for Carnap's position' (1991, p. 306). One could also say that he was much more interested in and gifted for a clear exposition of the different options we have at our disposal and the different choices we can make in philosophy, for example on so-called foundational questions in the philosophy of mathematics and the philosophy of science in general, than with a convincing explanation of the reasons for his own choice and the reasons a philosopher can find to choose in one way or another. He was obviously reluctant to enter really and seriously into philosophical controversies of any kind, at least in the usual sense of the term, and he thought that, once the questions have been formulated in an appropriate manner, i.e. as questions concerning the choice of the most convenient language for a determinate purpose, there will no longer be any real room for controversy. What would remain is only a peaceful confrontation between different choices that can be compared in an essentially pragmatic way. Adding the principle of tolerance to the thesis of metalogic, we should obtain as result a kind of perpetual peace in philosophy. The plurality of the philosophical answers and philosophical positions would, of course, subsist; but it would no longer be a problem, since nobody would be tempted to claim that his choice is the only possible one or to formulate absolute assertions of any kind.

We may remark that even on a question like the greatly controversial one, 'Is logic a matter of convention?', Carnap did not really wish to take sides and manifested the same conciliatory spirit as usual. When he wrote the *Logical Syntax of Language*, his point of view was purely syntactical, and the impression he gave was certainly that the rules of syntax have no responsibility of any kind to an antecedent meaning and can be chosen quite freely. But, after having realized that not only syntax but also semantics can be constructed in a completely formal and perfectly exact way, he thought that he had found a natural way to reconcile the conventionalistic point of view with the anti-conventionalistic one. It remains, of course, true, that the rules of a pure calculus C cannot be, in any sense, right or wrong and that we can choose them as we like, even if we take into account the fact that our system of syntactical rules has probably been designed to receive an interpretation and give rise to a language which we can use. As Carnap says in *Foundations of Logic and Mathematics*:

> We found the possibility – which we called the second method – of constructing a language system in such a way that first a Calculus C is established and then an interpretation is given by adding a semantical system S. Here we are free in choosing the rules of C. To be sure, the choice is not irrelevant; it depends upon C whether the interpretation can yield a rich language or only a poor one.
>
> We may find that a calculus we have chosen yields a language which is too poor or which in some other respect seems unsuitable for the purpose we have in mind. But there is no question of a calculus being right or wrong, true or false. A true interpretation is possible for any consistent calculus (and hence for any calculus of the usual kind, not containing rules for 'C-False'), however the rules may be chosen. (Carnap 1939a, p. 27)

To that extent, those who affirm the conventional character of logic, i.e. the possibility of a free choice of the logical rules of deduction, are right. But those who want to deny it may also be right. 'They are', Carnap says, 'equally right in what they mean, if not in what they say. They are right under a certain condition, which presumably, is tacitly assumed. The condition is that the "meanings" of the logical signs are given before the rules of deduction are formulated' (ibid.). The result of the confrontation is, therefore, the following:

> Logic or the rules of deduction (in our terminology, the syntactical rules of transformation) can be chosen arbitrarily and hence are conventional if they are taken as the basis of the construction of the language system and if the interpretation of the system is later superimposed. On the other hand, a system of logic is not a matter of choice, but either right or wrong, if an interpretation of the logical signs is given in advance. But even here, conventions are of a fundamental importance, for the basis on which logic is constructed, namely, the interpretation of the logical signs (e.g. by a determination of truth-conditions) can be freely chosen. (1939a, p. 28)

Thus, what remains of the initial debate seems to be only the question of whether we prefer to choose the syntactical rules first and then give to the syntactical system C an interpretation by adding to it a semantical system S or to choose the interpretation first and then look for syntactical rules of transformation that would be in agreement with the presupposed semantical rules. What is true in what the anti-conventionalists say is that, if we assume the semantical rules as given, and that is, for Carnap, the only thing we can mean when we say that the 'meaning' is given, 'we are [...] indeed bound in the choice of the rules in all essential respects' (ibid.). Thus, it is clear that we cannot expect Carnap to be ready to answer what could seem to be the properly philosophical question: are there independently existing meanings which determine what the semantical rules for the logical signs and maybe also for the signs of the language in general should be? The question has for him no real sense, since to say that meanings are already given can only mean that semantical rules have already been given. If by a 'factuality of meaning' we mean that there are facts concerning a world of meanings which could be given before the rules, there is, for Carnap, indeed no factuality of meaning. Our choices may depend on other choices, which we have previously made, but not on anything which could be said to be given in that sense.

## 2 Carnap and the way to avoid philosophical controversies

Even if there is little hope that a peaceful state of the kind some philosophers have dreamt of could ever be reached in their discipline, it is still possible for a philosopher like Carnap to avoid any kind of philosophical war, in the traditional sense. If asked whether some philosophers who do not accept his proposal concerning the real meaning and reference of philosophical propositions are saying something false or something meaningless, the Carnap of the *Logical Syntax of Language*, 'will', says Coffa, 'smile tolerantly and say, "Who am I to judge? I certainly don't understand what they are talking about, but I am no longer a dogmatist like Wittgenstein. Let each do as he chooses and let us all leave in peace"' (1991, p. 315). Carnap usually describes 'external' questions, as he calls them, i.e. questions concerning the choice of a linguistic framework, as non-cognitive in character and the answers which can be given to them as neither true nor false. He says that external questions, that is to say, questions of the form 'Shall we introduce such and such forms into our language?' are in the end not properly theoretical, but practical questions; they 'concern practical decisions rather than assertions' (1950a/1956, pp. 208 and 214). But that does not mean, for him, that they can be decided in an arbitrary way and that the answers cannot be discussed. Take, for example, the traditional question of the real existence of the thing world, as opposed to the world of sensations. When the question is translated into the formal mode of speech, it becomes, 'Shall we introduce into our language terms for physical things, and not only for sensations and classes of sensations?' And there are very good reasons to answer 'Yes'. Carnap is the first to admit that: 'The thing language in the customary form works indeed with a high degree of efficiency for most purposes of everyday life. This is a matter of fact, based on the content of our experiences' (ibid., p. 208). However, we must not go so far as to treat the matters of fact which are involved in the question as 'confirming evidence for the reality of the thing world' (ibid.).

The reason for that is quite clear. Different linguistic frameworks can be more or less useful or efficient for different purposes. But to speak of usefulness or efficiency for all purposes and in an absolute sense has, for Carnap, no real sense. Moreover, if the question is formulated, as it should be, in the following form: 'Are our experiences such that the use of linguistic expressions designating physical things will be expedient and fruitful?', 'this', Carnap says, 'is a theoretical question of a factual, empirical nature. But it concerns a matter of degree; therefore a formulation in the form "real or not" would be inadequate' (1950a/1956, p. 213). According to Bryan Norton, 'The central thrust of the principle of tolerance is to emphasize that all judgments concerning linguistic frameworks must take place in a particular context with a clearly stated purpose' (1977, p. 139). And that, of course, is particularly true of judgements concerning the utility of a linguistic framework and the superiority and preferability, in whatever sense that has to be understood, which one may want to attribute to it.

In answer to a suggestion of Bar-Hillel, who had remarked that 'discussing the utility of a proposal is essentially the same as discussing the truth of the assertion that this proposal is useful' (1963, p. 536), so that philosophical sentences could perhaps, in the end, be seen as being really theoretical assertions concerning the factual question of the utility of a linguistic framework, Carnap said:

> Bar-Hillel suggests not only to *replace* ontological theses of the existence of certain kinds of entities by a discussion of practical questions concerning the choice of forms of language, but rather to *interpret* those theses as assertions of the expediency of corresponding language forms for certain purposes. It is true that this procedure would have the advantage that the allegedly theoretical theses of ontology would be interpreted as genuine theoretical theses. However, I still have the feeling that this re-interpretation deviates too much from the interpretation which the philosophers themselves actually had in mind. (Carnap 1963b, pp. 941–2)

Thus, it seems that Carnap's position was the following one: we can propose, at least to those who are likely to be sensitive to the interest of a suggestion of that kind, to replace the traditional ontological discussions with a discussion of practical questions concerning the choice of forms of language; but that is not exactly the same as proposing an interpretation of the philosophical theses in question, and, if the proposal was intended as an interpretation, the objection would be that it is too far from the interpretation that the philosophers who formulate them have in mind. The discussion of the philosophical theses will, it is true, be replaced by the (much more intelligible and fruitful) discussion of something which might seem at first sight very remote from what they say and, consequently, of what their authors had in mind; but that is not an inconvenience, since what is at stake is not an interpretation.

Carnap, in the *Logical Syntax*, used as one of his central examples of situations in which the principle of tolerance has to be applied the controversy, in the philosophy of mathematics, between intuitionism or, more generally, constructivism, and classicism. He constructed two languages, the weaker Language I and the more powerful Language II, in relation to these two options, and he recommended (and practised) complete tolerance with respect to the choice between Language I and II and to every choice of the same kind. Some of the prohibitions which have been formulated, for example in the philosophy of mathematics, have been historically useful in that they have permitted important differences to be emphasized. But there is, Carnap thinks, no reason to continue to set up prohibitions, since prohibitions can now be replaced by 'a definitional differentiation' and the simultaneous investigation of language-forms of different kinds constructed for different purposes (LSL, p. 51).

The reason for Carnap's attitude is easy to find. It is the recognition of the fact that a language like his Language I, albeit weaker, may still be superior and preferable for certain purposes, while Language II is better for other purposes. A serious philosopher, according to Carnap, should, therefore, never ask questions like 'Is it the point of view of the intuitionists or the point of view of the classical mathematician which is the right one?' The question is not 'Who tells the truth?', but 'Who makes the better choice and the better proposal?' But even the latter question cannot have the kind of absolute answer a philosopher would probably be still waiting for. We should also refrain, in philosophy, from asking questions like 'Is Language II more suitable and to be preferred to Language I?' For, such completely decontextualized questions, even if they have the form of syntactical questions concerning language and linguistic expressions, are in reality meaningless. The only thing of which we can speak meaningfully in such matters is the suitability of a given language-form or, as Carnap later said, a given linguistic framework for a certain task. To think otherwise would be, for him, to keep to a metaphysical point of view and to continue to think in the metaphysical way. What is metaphysical is the belief that we can hope for and have to look for yes or no answers to questions like 'Is the thing world real?' or 'Is intuitionistic mathematics the true mathematics?'

But there is, it seems, a fundamental question, which it is impossible to avoid: if we accept Carnap's point of view, what remains exactly of philosophical controversies and even of philosophical discussions, as they have been traditionally conceived? Is there still anything to be decided in philosophy by philosophical means? And is there a real necessity, for philosophy understood in the Carnapian way, to try to decide it? It is not possible to decide directly a philosophical question like 'Is the thing world real?', as it is usually formulated, since it has no real meaning and we have, to say the least, no clear idea of what is really asked and what is really at stake. But once the question has been given an acceptable and intelligible form by an appropriate translation into the formal mode of speech, a Carnapian philosopher would no longer feel any need to try to decide it.

Warren Goldfarb describes in the following way Carnap's notion of a language or, according to the terminology he used later, a linguistic framework in the *Logical Syntax*:

> A linguistic framework is given by the rules for formation of sentences together with the specification of the logical relations of consequence and contradiction among sentences. The fixing of these logical relations is a precondition for rational inquiry and discourse. There are many alternative frameworks, many different logics of inference and inquiry. Since justification can proceed only grounded in the logical relations of a particular framework, justification is an intraframework notion. Thus there can be no question of justifying one framework over another. Carnap voices this pluralistic standpoint in his Principle of Tolerance: '*In logic there are no morals*. Every one is at liberty to build up his own logic, i.e. his own form of language, as he wishes. All that is required of him is that ... he must state his methods clearly and give syntactical rules instead of philosophical arguments' (LSL, §17). Now Carnap so defines the notion of 'analytic' that a sentence is analytic in a linguistic framework if it is a consequence of every sentence. In calling logic and mathematics analytic, Carnap is thus saying that they consist of framework-truths: sentences that any user of the linguistic framework must accept, just by dint of his being a user of that linguistic framework: they are simply consequences of the decision to adopt one rather than another linguistic framework. (Goldfarb 1996, p. 225)

Carnap says that we must give syntactical rules instead of philosophical arguments. But what about philosophical arguments for or against the choice of a particular linguistic framework? Is it still possible to speak in any sense of arguments of that kind? Even Carnap could, it seems, hardly accept completely the idea that justification can be only a purely intraframework notion. Some commentators have suggested that his views in 'Empiricism, Semantics, and Ontology' are equivalent to the view that the acceptance or refusal of any external or philosophical position 'will in the end be *wholly* grounded on a free choice, a nonrational leap of faith' (D. Goldstick, quoted by Norton [1977, p. 132]). But he clearly thought that properly formulated external questions have a kind of rationality and can, at least to some extent, be discussed rationally. However, it is tempting to draw from his pluralistic and relativistic attitude concerning the choice between different possible linguistic frameworks the conclusion that what would be rational, in matters of that kind, is rather not to feel obliged to choose, and that, if we nevertheless make a choice, it cannot but be irrational or at least non-rational.

Warren Goldfarb is certainly right when he emphasizes that the syntactical interpretation and reconstruction which Carnap proposes for mathematical propositions is not designed to answer a question of justification:

> It should be clear [...] that Carnap does not view the reduction of mathematics to syntax as providing a justification for mathematics; the identification of mathematical truths as framework-truths is not meant to legitimize them. Carnap could allow that, while mathematical truths are the result of syntactic rules, our recognition of particular truths are the result of syntactical rules, our recognition of particular truths, or our trusting any particular formulation of what can be inferred from given syntactical rules requires more mathematics or different mathematics than that which those rules yield. (Of course, he would also assert, the additional or different mathematics we use is also the upshot of syntactical rules, albeit different ones.) In short, Carnap is not taking the clarification of the status of mathematics contained in *Logical Syntax* to be addressing traditional foundational issues. Those issues are addressed in another way, for they are transformed into questions of what can be done inside various linguistic frameworks or what sort of frameworks are better for one or another purpose. What remains of 'foundations of mathematics' is treated by describing, analyzing and comparing different frameworks. (1996, pp. 228–9)

That seems to me to be quite true and constitutes the main reason why Carnap would not have been much impressed by an argument which was used by Gödel in order to refute the thesis according to which mathematical propositions have to be assimilated to rules of logical syntax. If mathematics is reduced to a system of linguistic conventions, we must at least be able to prove that that system of conventions is a possible one, i.e. is not inconsistent. For a rule can be called syntactical only if it can be known beforehand that it does not imply the truth or falsehood of any 'factual' sentence. And a rule or a system of rules which is inconsistent would clearly not satisfy this requirement, since it would entail any sentence, true or false and factual or not factual. That means that, if we want to propose a system of conventions concerning a given linguistic framework as a plausible candidate for the representation of mathematics as it is, we must be able at least to establish that the system in question is not inconsistent. But Gödel's second incompleteness theorem states that, in order to prove that the system is consistent, we will have to use mathematical means which cannot be represented in the system itself. However, the objection can be decisive only if the problem Carnap wanted to solve was one of foundation or of justification. But that was not his problem and he did not, for example, worry about the fact that we may have to use in the language of syntax non-finitary (in his language, 'indefinite') terms that cannot be defined in the language we consider. For that too – the question of whether indefinite terms can or cannot be admitted in syntax – is a matter to be decided by convention (LSL, p. 165) and subjected to the principle of tolerance.

## 3 Carnap and the way of tolerance in philosophy

According to Goldfarb, 'In *Logical Syntax*, Carnap no longer takes there to be any questions about logic and mathematics that are foundational in the traditional sense. He is simply no longer addressing the issues that concerned Kant, Frege, Russell, Hilbert, or even the Carnap of "Logizistische Grundlegung"' (1996, p. 229). That is the reason why there can be no real debate with Gödel, who is asking the questions in traditional, that is to say, foundational or epistemological terms, and sees them as requiring yes or no answers. A question of that kind, for which we must be able to provide a yes or no answer, is 'Do mathematical sentences have a content?' And Gödel thinks that a positive answer can be given to it, at least in the sense that the thesis that mathematical sentences are contentless can be refuted and has been refuted. He considers that the theses of the philosophers who treat mathematical truths as linguistic conventions 'are refutable, as far as any philosophical assertion can be refutable in the present state of knowledge' (quoted by Warren Goldfarb [1995, p. 325]). For him, the real progress in philosophy, at least in a domain that can already be treated in an exact way, like the philosophy of mathematics, would be to succeed in giving philosophical assertions a form which is sufficiently clear and precise to make them not only disputable, but also testable and refutable. And the evolution of the search for the foundations of mathematics has already produced results that could lead to the refutation of some philosophical theses.

But that is almost exactly the contrary of the result that can be obtained if one adopts the point of view of the *Logical Syntax*. For Carnap, we will never be and we should not try to be in a position that would permit us to refute, and, as a consequence of that, to exclude, a philosophical conception about mathematics. '*It is*', he says, '*not our business to set up prohibitions, but to arrive at conventions*' (LSL, p. 51). But it is difficult to see how we could succeed in giving a refutation of some philosophical thesis and at the same time abstain completely from formulating a prohibition of any kind. Everyone remains of course free to accept even a proposition which has been shown to be false. But it does not seem to be possible to refute a philosophical position without trying to persuade those who hold it that it is does not, in reality, represent a possible option and has, therefore, to be excluded. But that is, it seems, already more than we can, for Carnap, feel authorized to do and hope to be able to do in philosophy.

As I have said, Coffa considers that, 'if Carnap's application of his linguistic techniques in LSL is to have any relevance to the foundational problems that others were debating, it must be because behind the first-level semantic conventionalism there is a second-level factualism that poses a genuine challenge to all other philosophies of mathematics' (1991, p. 322). But perhaps Carnap had no wish to pose a challenge to people who have preferences and make choices different from his own in the philosophy of mathematics. We may even wonder whether it would not be possible to generalize Goldfarb's statement and say that Carnap was simply no longer addressing philosophical questions in the traditional sense. It is the use of the material mode of speech which is responsible for the ignorance of what he calls 'a very essential point to keep in mind', i.e. '*the relativity of all philosophical theses to language*' (Carnap 1935a, p. 78). But, as we have seen, once the translation into the formal mode of speech has made obvious the incompleteness which affects the philosophical theses, because of an unnoticed want of reference to language and, more precisely, to a determinate language, it is really difficult to know what finally subsists of the initial controversy.

It certainly, as Carnap says, 'becomes clear and exact' (ibid.) and, in certain cases, it may even simply disappear. That is exactly what happens, according to him, in the case of the two philosophical assertions 'Numbers are classes of classes', and 'No, numbers are primitive objects, independent elements'. If the language-system of Peano is called $L_1$, and that of Russell $L_2$, the two sentences, formulated in the syntactical mode, will be completed as follows: 'In $L_1$ numerical expressions are elementary expressions' and 'In $L_2$ numerical expressions are class expressions of the second order'. 'Now', Carnap says, 'these assertions are compatible with each other and both are true: the controversy has ceased to exist' (ibid., p. 77). But, in most cases, things are not so simple. Even when the controversy has become clear and exact, it may be still be difficult, Carnap tells us, to decide which side is right (ibid., p. 78). But it could also appear that it is now much less important to decide and even that it is no longer possible to speak seriously of a side which is right.

Carnap does not say that the sentences in the material mode of speech are themselves necessarily pseudo-theses or without sense. The problem with them is only that they can easily mislead us into asking questions and formulating statements which seem to be meaningful, but are not. Carnap says:

> For instance, in the material mode of speech we speak about numbers instead of numerical expressions. That is not in itself bad or incorrect, but it leads us into the temptation to raise questions as to the real essence of numbers, such as the philosophical questions whether numbers are real objects or ideal objects, whether they are extramental or intramental, whether they are objects-in-themselves or merely intentional objects of thinking, and the like. I do not know how such questions could be translated into the formal mode or into any other unambiguous and clear mode; and I doubt whether the philosophers themselves who are dealing with them are able to give us any such precise formulation. Therefore it seems to me that these questions are metaphysical pseudo-questions. (LSL, p. 79)

When philosophical sentences speaking of objects of some kind in the material mode of speech are translated into the formal mode of speech, they are replaced by correlated sentences of the formal mode of speech making corresponding assertions about the designations of these objects. And one of the main advantages of the formal mode of speech is that, for a lot of obscure and seemingly irresoluble questions that could be asked about the objects themselves, for example whether they are real or ideal, mental or non-mental, existing in themselves or only intentionally, there are simply no corresponding questions that can be meaningfully asked about the expressions that serve to designate them. There is obviously no property of expressions expressible in syntactical terms corresponding to properties of objects like real or ideal, mental or non-mental, etc. Carnap calls 'quasi-syntactical sentence' a sentence which attributes to an object a 'quasi-syntactical property', i.e. a property 'which is, so to speak, disguised as an object-property, but which, according to its meaning, is of a syntactical character' (LSL, p. 234). And properties like real or ideal, mental or non-mental etc., are clearly not quasi-syntactical. Or, in any case, it is up to those who want to state and discuss assertions like 'Numbers are not real but ideal objects', to explain how the properties they speak of could be interpreted as quasi-syntactical and replaced by correlated syntactical properties of their designations. If philosophers want to discuss, they must begin by giving to their sentences a form that makes discussion possible. Carnap is, on that point, quite clear: 'If one partner in a philosophical discussion cannot or will not give a translation of his thesis into the formal mode, or he will not state to which language-system his thesis refers, then the other will be well-advised to refuse the debate, because the thesis of his opponent is incomplete, and discussion would lead to nothing but empty wrangling' (1935a, pp. 80–1). There would still not be, for Carnap, any prohibition formulated against a philosophical statement of any kind, since that means only that in some cases and maybe even in most cases we could feel authorized simply to ignore what the philosophers assert or negate. But there is, to say the least, a little problem with the use Carnap makes of the problematic notion of meaning when he says that there are properties which seem to be object-properties, but according to their meaning, are of a syntactical character. The crucial question is, of course, to know whether there is any chance that the philosophers will accept the analysis Carnap gives of their meaning and recognize the meaning of the theses they have tried to state, after they have been translated into the formal mode of speech.

It is remarkable to see Carnap immediately apply his apparatus in a way that permits him to show that a dispute between logicists and formalists concerning the nature of numbers boils down simply to a pseudo-issue. The logicists hold that numbers are classes of classes of things, while the formalists assert that they are given individual entities. We have seen how the apparent conflict could, according to Carnap, be definitively settled, leaving only the question of the pragmatic advantages of two different languages to be discussed. And in some cases it is not even necessary to refer to two different language-systems in order to make the two philosophical theses compatible with one another. Carnap suggests that the positivist philosopher, who says: 'A thing is a complex of sense-data', and his realist adversary, who replies: 'No, a thing is a complex of physical matter', can easily be reconciled. For, when their respective statements are translated into the formal mode of speech, it appears that both of them were saying something true, already 'in relation to our general language' (ibid., p. 82). What the positivist asserted is, in effect, that: 'Every sentence containing a thing-designation is equipollent with a class of sentences which contain no thing-designations, but sense-data-designations', and what the realist replied was: 'Every sentence containing a thing-designation is equipollent with a sentence containing no thing-designation, but space-time coordinates and physical functions'. Both of them are right and the reason why the theses in the original formulation seemed incompatible, is, observes Carnap, that they seemed to concern the essence of things, 'both of them having', he says, 'the form: "A thing is such and such"' (ibid.). It would, therefore, be better simply to avoid a form of that kind, which creates a completely misleading appearance of incompatibility between sentences which can in reality very well be simultaneously true. To say that would be, of course, of little help and little use to a philosopher who, like Heidegger, thinks that what philosophy, as opposed to the sciences in general, is striving for is precisely 'essential knowledge'. But Carnap does not see any possibility of arguing in an interesting and fruitful way with a philosopher who maintains that philosophical sentences are genuine object-sentences, dealing with an extralinguistic reality.

It seems, therefore, that the most difficult and the most decisive aspect of the problem is: 'What are exactly philosophical sentences, at least philosophical sentences of the redeemable kind, i.e. to which we may hope to be able to give a meaning, speaking of?' or 'What are they about? How can we hope to solve a disagreement that cannot but be itself philosophical concerning that very question?' Thus, from the Carnapian point of view, all depends, in a sense, on what the syntactical approach can retain of the question of 'aboutness' or of what comes nearest to it, once we have renounced speaking, in the material mode, of a relation between our linguistic expressions and an extralinguistic reality. Carnap, in his pre-semantical period, tells us that the sentence 'This book treats of Africa' is a sentence of the material mode, which can be translated into the formal statement 'This book contains the word "Africa"'. And he adds: 'Similarly, to the material mode belong all these sentences which assert that a certain sentence or treatise or theory or science *deals with* such and such objects, or *describes* or *asserts* such and such facts or states or events; or that a certain word or expression *designates* or *signifies* or *means* such and such an object' (ibid., p. 71).

Among the many examples he gives, in the *Logical Syntax of Language*, of the way problematic sentences of the most different kinds in the material mode of speech could be translated into unproblematic sentences of the formal mode, one of the most astonishing and disconcerting is certainly the treatment he proposes to apply to sentences like 'Yesterday's lecture treated of Babylon'. The correlated syntactical sentence that we can use instead of this one is: 'In yesterday's lecture the word "Babylon" (or a synonymous designation) occurred' (LSL, p. 289). Even if we grant to Carnap that a satisfactory definition of synonymy or identity of content between two linguistic expressions can be given in purely syntactical terms, we may doubt seriously the possibility of replacing the first sentence with the second one. A book could contain an occurrence of the word 'metaphysics', but only in order to tell the reader that the author has no intention to treat of metaphysics, and it might be indeed a book that does not deal with metaphysics. Conversely, somebody who would say of a book that, although it does not contain the word 'metaphysics' or any other word or expression synonymous with it, it nevertheless treats of metaphysics could very well tell the truth in certain cases.

But Carnap has, on questions of that kind, no real choice. For the only means to save sentences which express a relation of designation, that is to say, those in which occur expressions such as 'treats of', 'speaks about', 'means', 'signifies', 'names', 'is a name for', 'designates' and the like is, for him, to treat them as quasi-syntactical sentences in the material mode of speech, that is to say, sentences for which there is a possibility of translation into the formal mode, even if it is probably not a completely satisfactory one. And it is difficult to see what other equivalent we could find for them in the formal mode, than the one he proposes. For what could be the purely syntactical property that belongs to the designation 'Babylon' whenever the thing designated is a thing to which reference is made in a sentence, a treatise, a theory, or a science? Carnap considers that sentences which assert something about the *meaning*, *content*, or *sense* of sentences or linguistic expressions cannot be object-sentences. They are what he calls in the *Logical Syntax* 'logical sentences', as opposed to 'object-sentences', but they have the form of pseudo-object-sentences. Carnap gives as a reason for that the fact that the sentence 'Yesterday's lecture was about Babylon' only appears to say something about the town Babylon. In reality, in spite of the fact that it contains the name of the town, it says something only about yesterday's lecture and the word 'Babylon': 'For our knowledge of the properties of the town of Babylon, it does not matter whether [the sentence] is true or false' (LSL, pp. 285–6).

One could be tempted to object that nothing, from a logical point of view, can prevent us from saying that the sentence 'Yesterday's lecture was about Babylon' could be analysed not only as asserting something about yesterday's lecture, but also as asserting something about the town Babylon itself, i.e. that it is an object to which reference is made in yesterday's lecture or of which yesterday's lecture spoke, even if to know that would indeed add nothing really significant to our knowledge of the properties of the town. But that is not the most important aspect of the problem. For the possibility Carnap wants to exclude is in reality that the sentence is an object-sentence dealing not exactly with the town Babylon, but rather with an object called the content, the meaning, or the sense of yesterday's lecture. A supposititious object of that kind is for him only a pseudo-object. But it is curious to see him using as his first argument for that the fact that the sentence, contrary to what it seems to do, does not treat of the town Babylon.

In answer to the question of why he repeatedly proposes to translate sentences which are formulated in the material mode of speech into the formal mode, Carnap says that the syntactical character of these sentences is disguised: 'We are deceived – as we have seen – as to their real subject matter' (1935a, p. 76). And most philosophical sentences happen to be formulated in a mode of speech that typically deceives us as to their real subject matter. But there will be, it seems, a problem with a sentence purporting to assert that the real subject-matter of a given sentence is not what it seems to be, but something completely different. For if the only way to assert the fact that the sentence S speaks of an object *a* is to say that it contains a designation of *a*, it will be simply impossible to find an acceptable syntactical correlate for the sentence in question.

In LSL, Carnap distinguishes between object-questions that are concerned with supposititious objects which are not to be found in the object-domains of the sciences (he gives as examples of objects of that kind the thing in itself, the absolute, the transcendental, the objective idea, the ultimate cause of the world, non-being, and things like values, absolute norms, the categorical imperative, etc.) and questions concerned with things which also occur in the empirical sciences (such as mankind, society, language, history, economics, nature, space and time, causality, etc.). As regards those object-questions whose objects do not appear in the exact sciences, the situation is clear: the logical analysis of the philosophical problems has shown that they are pseudo-problems. Once they have been eliminated, there remain, apart from the questions of the individual sciences, only the questions of the logical analysis of science, of its sentences, terms, concepts, theories, etc. That complex of questions is called by Carnap the *logic of science* and his suggestion is that the logic of science has to take the place of what he calls 'the inextricable tangle of problems which is known as philosophy' (LSL, p. 279).

But he does not present that as a thesis which has been established. Somebody who shares the anti-metaphysical point of view will conclude that all philosophical problems that have any meaning belong indeed to syntax. But if a philosopher does not accept that conception, he will be free to interpret what Carnap states as meaning only that 'the problems of that part of philosophy which is neither metaphysical nor concerned with values and norms are syntactical' (ibid., p. 280). Since nobody could be forced to adopt the anti-metaphysical stance, nobody could be forced to accept the idea that the only philosophical problems that have a real meaning belong to syntax. Everybody remains free to continue to believe that there are really metaphysical problems, i.e. problems that have a metaphysical meaning, which cannot be interpreted as a syntactical one. As we can see, tolerance, for Carnap, was not simply a word or a motto, which can be forgotten as soon as it has been formulated. He really practised himself, in philosophy as well as in logic, what he had preached and his conception of tolerance went surprisingly far. He really did not want ever to appear as trying to formulate prohibitions of any kind and he was very anxious to present the anti-metaphysical programme itself in such a way that it would not sound in any way like a prohibition. We may, of course, admire him for his radically anti-dogmatic attitude and the degree to which he was ready to use tolerance. But we may also think that even tolerance, in philosophy, can sometimes go too far and that the distance from complete tolerance to philosophical indifference is perhaps not very great.

# 7

# The Analysis of Philosophy in *Logical Syntax*: Carnap's Critique and His Attempt at a Reconstruction

*Pierre Wagner*

Like other members of the Vienna Circle, Carnap criticized traditional philosophy for its lack of clarity and precision, and he promoted a style of thinking more akin to scientific practice than to poetry or other forms of art. According to this line of thought, the everlasting struggles between metaphysical systems based on personal worldviews or on original intuitions should give way to a collective endeavour which may well be inspired by emotions and feelings but has ultimately to be given 'a purely empirical – rational justification' (Carnap 1928a/2003, p. xvii). In the case of Carnap and the other members of the so-called 'left wing' of the Vienna Circle, such commitment to a scientific conception of the world went far beyond the limits of academic disputes and took the form of an intellectual engagement reminiscent of the Enlightenment, which aimed at nothing less than the 'conscious re-shaping of life' (Carnap, Hahn, Neurath 1929/1973, p. 305) and at social progress, although Carnap himself did not frequently make such pronouncements in his philosophical writings.[1] *The Logical Syntax of Language* (LSL from now on) should no doubt be considered as inspired by the same spirit, although this particular aspect will probably not be the first thing to strike the reader, especially if she has no prior knowledge of the historical context in which it was written.

Carnap's most well-known criticism of traditional philosophy consists of his rejection of metaphysics and, more specifically, of the analysis of metaphysical propositions as nonsense. Carnap was not the only one in the Vienna Circle who took up Wittgenstein's radical criticism of philosophical propositions in the *Tractatus*:

> Most of the propositions and questions to be found in philosophical works are not false but nonsensical. [...] Most of the propositions and questions of philosophers arise from our failure to understand the logic of our language.

> (They belong to the same class as the question whether the good is more or less identical than the beautiful.) (Wittgenstein 1921/1961, 4.003)

In 1932, Carnap's own justification for dismissing metaphysics was most memorably expounded in 'The Elimination of Metaphysics through Logical Analysis of Language':

> In the domain of *metaphysics*, including all philosophy of value and normative theory, logical analysis yields the negative result *that the alleged statements in this domain are entirely meaningless*. Therewith a radical elimination of metaphysics is attained, which was not yet possible from the earlier antimetaphysical standpoints. (1932a/1959, pp. 60–1)

This point is made more precise when Carnap further expounds two reasons why the alleged statements of metaphysics are 'pseudo-statements':

> either they contain a word which is erroneously believed to have meaning, or the constituent words are meaningful, yet are put together in a counter-syntactical way, so that they do not yield a meaningful statement. (ibid., p. 61)

In the second case, the pseudo-statements violate the logical syntax of language. Such criticism clearly depends on a theory of meaning, and Carnap does not fail to provide one in the following sections of the paper, for both words and sentences. This takes the form of an uninhibited verificationist criterion of meaning (for the object-language): on the one hand, 'a word is significant only if the sentences in which it may occur are reducible to protocol sentences' (ibid., p. 63) and on the other hand, 'the meaning of a statement lies in the method of its verification' (ibid., p. 76). Carnap concludes that the pseudo-statements of metaphysics do not have any *cognitive* meaning – they 'do not serve for the *description of states of affairs*' (ibid., p. 78) – although they may have an *expressive* meaning in so far as they are capable of arousing feelings and emotions and thus of giving 'the expression of the general attitude of a person towards life' (ibid., p. 78). Metaphysicians often pretend to express authentic knowledge – indeed, some kind of special and important knowledge deemed superior to mere *scientific* knowledge. As a result their pseudo-statements breed confusion and should be eliminated not only as cognitively meaningless but also as especially misleading.

Although this line of argument encapsulates the Carnapian criticism of philosophy which is probably most frequently referred to, it is actually endorsed neither in the *Aufbau* nor in LSL and, as a matter of fact, Carnap maintained it only for a short period of time. In the *Aufbau*, Carnap did reject 'intuitive metaphysics' as being non-rational and unscientific, but not on the grounds of its meaninglessness. As Michael Friedman notes, the basis for Carnap's anti-metaphysical attitude in the *Aufbau* is neither the principle of verifiability nor empiricist reductionism (2007, p. 147). Moreover:

> Carnap, in the *Aufbau*, is by no means uninterested in the traditional metaphysical dispute between realism and idealism. On the contrary, he devotes considerable ingenuity and philosophical imagination to crafting logical reconstructions of these positions which capture what he takes to be correct and uncontroversial in them. (Friedman 2007, p. 150)

Carnap was looking for a way to turn the endless philosophical disputes into a scientific enterprise and he regarded the theory of constitution expounded in the *Aufbau* as 'the "neutral foundation" common to *all* epistemological tendencies' (ibid., p. 149).

On the other hand, *after* writing 'The Elimination of Metaphysics through Logical Analysis of Language', Carnap soon gave up the verificationist criterion of meaning in its strict form and his line of argument against philosophy in LSL is not grounded in this criterion either. To be sure, Carnap still maintains that the questions concerned with such pseudo-objects as 'the thing-in-itself, the absolute, the transcendental, the objective idea, the ultimate cause of the world, non-being, and such things as values, absolute norms, the categorical imperative, and so on' (LSL, p. 278) give rise to meaningless pseudo-statements which express no theoretical sense:

> Metaphysical propositions are neither true nor false, because they assert nothing, they contain neither knowledge nor error, they lie completely outside the field of knowledge, of theory, outside the discussion of truth or falsehood. But they are, like laughing, lyrics, and music, expressive. (1935a, p. 29)

However, this view no longer depends on strict verificationism, and it is more fully expounded in 'popular' expositions such as Carnap (1935a) than in LSL.

Carnap does occasionally refer to 'the anti-metaphysical attitude represented by the Vienna Circle' in LSL, but he then adds that this attitude 'will not, however, appear in this book either as an assumption or as a thesis' (LSL, p. 8). In Part V, when he elaborates a critique – and, as we shall see, a reconstruction – of philosophy based on the method of logical syntax, far from rejecting philosophical sentences as meaningless, he argues that many sentences which are to be found in philosophical writings do qualify as meaningful, although an understanding of their actual place in the system of science is usually obscured by the inappropriate way in which they are formulated. Only their translation into a more formal mode of speech is likely to clarify their real nature and make it clear that 'the problems of that part of philosophy which is neither metaphysical nor concerned with values and norms are *syntactical*' (LSL, p. 280, emphasis added). This claim, made in the light of logical analysis, obviously requires clarification. In particular, we need to know what exactly is meant by such a diagnosis and which arguments are put forward to justify it.

What Carnap points out and criticizes is not the meaninglessness of philosophy but the wrong understanding we have of its relationships with science and the confusion which results from its mixed character: 'the logical analysis of philosophical problems shows them to vary greatly in character' (LSL, p. 278). Philosophy is a 'tangle of problems' (LSL, p. 279) which – through logical analysis – separate out into three categories of questions: meaningless metaphysical questions which should be eliminated, object-questions of the special factual sciences which should be dealt with using empirical methods, and logical questions dealing with the language of science. The main point is that once metaphysical pseudo-sentences are eliminated as meaningless, the 'remaining philosophical questions are logical ones' (LSL, p. 279) because there is nothing such as an investigation of the *objects* of the individual sciences from a non-scientific, purely philosophical viewpoint:

> Apart from the questions of the individual sciences, only the questions of the logical analysis of science, of its sentences, terms, concepts, theories, etc. are left as genuine scientific questions. We shall call this complex of questions the *logic of science*. (LSL, p. 279)

So the upshot of logical analysis is not the elimination of philosophy but a better understanding of its real nature, to wit its *logical* nature. Although this is reminiscent of Russell's famous claim to the effect that logic is the essence of philosophy (Russell 1914, ch. 2), Carnap's point is made in a context completely foreign to Russell's understanding of logic: logic is interpreted as syntax and in LSL the fundamental distinction between object-language and syntax-language is taken for granted. Carnap's own claim is that philosophical questions really are syntactical questions about the language of science – regarded as an object-language – being raised in a metalanguage. Hence his pronouncement in the foreword of LSL: '*philosophy is to be replaced by the logic of science* – that is to say, by the logical analysis of the concepts and sentences of the sciences, for *the logic of science is nothing other than the logical syntax of the language of science*' (LSL, p. xiii).

In the postscript to the English translation of (1932a), Carnap names a few authors who qualify as typical targets for his attacks on metaphysics: Fichte, Schelling, Hegel, Bergson, and Heidegger (1932a/1959, p. 80).[2] In Part V of LSL, when he analyses examples of sentences formulated in a misleading way, the authors he has in mind are not at all the same. Many of the philosophical sentences analysed in this part of the book are from Wittgenstein's *Tractatus*, some are from mathematicians such as Kronecker,

Weyl, and Brouwer, and others could have been written by Russell, Schlick, or by other authors whose writings belong to what Carnap calls 'the logic of science' in §72:

> The term '*logic of science*' will be understood by us in a very wide sense, namely, as meaning the domain of all the questions which are usually designated as pure and applied logic, as the logical analysis of the special sciences or of science as a whole, as epistemology, as problems of foundations, and the like (in so far as these questions are free from metaphysics and from all reference to norms, values, transcendentals, etc.) To give a concrete illustration we assign the following investigations (with very few exceptions) to the logic of science: the works of Russell, Hilbert, Brouwer, and their pupils, the works of the Warsaw logicians, of the Harvard logicians, of Reichenbach's Circle, of the Vienna Circle centring around Schlick [...] (LSL, pp. 280–1)

At first sight, it is difficult to see how such a wide sense can be compatible with Carnap's other characterization of the logic of science given on the same page: 'the logic of science is the syntax of the language of science'. Given Carnap's understanding of 'syntax' in LSL,[3] few of the authors cited as illustrations of the logic of science 'in the wide sense' would actually regard their own works as coming under 'the syntax of the language of science'. Carnap is aware of that, but his point is that the formulations actually used by these authors in their works often tend to obscure their syntactical character, which would become clear if they were translated into the 'formal mode of speech'. One of Carnap's most frequently used examples of a sentence in the material mode of speech is '5 is a number'.[4] Whereas this sentence apparently expresses a property of some mathematical *object*, its translation in the formal mode of speech as '"5" is a number-word' expresses the property of a *word* and thus belongs to the logical syntax of some object-language. Unless we carefully distinguish these two sentences, we are in danger of blurring the fundamental difference between object-questions and logical questions.[5] To be sure, the logic of science 'in the wide sense' includes many works which actually do not take this distinction into consideration. According to Carnap, however, their translation into the formal mode of speech makes it clear that they essentially contain syntactical sentences and thus belong to the logic of science in the narrow sense. Properly understood, the philosophical sentences which are not meaningless can be clarified as coming under the logical syntax of the language of science.

We will not realize how far-reaching this diagnosis actually is unless we notice that it holds not only for those writings which obviously bear on logical matters: it also implies that any meaningful philosophical issue is expressible as a logical, i.e. a syntactical, question. In the framework of the logic of science, philosophical issues lose their absolutist character and become relative to language. This is what makes Carnap's analysis of philosophy constructive: far from being purely critical, he also attempts a reconstruction of philosophy in a new framework.[6]

In one of the three papers he read at the Paris Congress in September 1935, Carnap takes up again the idea of philosophy as a 'tangle of problems' and applies it to the special case of the theory of knowledge: 'It seems to me that in the form it has had so far, *the theory of knowledge* is an *unclear mixture of psychological and logical components*' (1936a, p. 36, my translation). The target of this criticism is then made more explicit: 'This holds also for the works of our Circle, without excluding my own former works' (ibid.). What Carnap has in mind here are writings such as Schlick's 1934 paper 'On the Foundation of Knowledge' (to which he alludes in the following lines), the *Aufbau* (1928a), and his *Pseudoproblems in Philosophy* (1928b). In these works (in which classical issues of the theory of knowledge play a prominent role), the authors do not distinguish the object-language and the syntax-language and fail to realize how the formal mode of speech would clarify the status of their philosophical discourse.[7] The translation suggested by Carnap from the material mode of speech into the formal mode of speech is not meant to introduce a new method or a new domain of investigation but to make us aware of the logical character of what is usually called 'theory of knowledge':

> When I say that the logic of science takes the place of the theory of knowledge, I do not thereby propose a new method. It seems to me, rather, that even in the works we have done so far, the non-psychological questions were questions belonging to logical syntax. [...] So all we want is that we now become aware of what we have always been doing already. (1936a, p. 37, my translation)

Whereas in Carnap (1936a) this remark holds for the special case of the theory of knowledge, in LSL, Carnap calls for a reinterpretation of the propositions of the logic of science 'in the wide sense' and of the propositions of philosophy in general. At the same time, he clarifies his own way out of a quandary which results from the general stance taken by the logical positivists. He thus puts forward a programme for scientific philosophy which appears to be at variance with the orientation taken by some of his friends in the Vienna Circle. In the mid-thirties, there were several issues on which Carnap and Schlick had opposed views although they usually tried to avoid making their disagreements public.[8] The function of philosophical propositions with respect to scientific ones was one of the disputed questions.

In the twenties, the logical empiricists had arrived at an analysis of scientific and philosophical propositions which resulted in the rejection of metaphysics and in a clear-cut distinction between the tautological propositions of logic and mathematics on the one hand and the synthetic propositions of empirical science on the other hand. The rejection of synthetic knowledge a priori was regarded as 'the basic thesis of modern empiricism' (Carnap, Hahn, Neurath 1929/1973, p. 308). This analysis, however, also questioned the logical positivists' own philosophical stance and they soon realized they had to face a serious objection that Carnap states in the following way:

> We take the view, expressed already by Hume, that besides logico-mathematical tautologies (analytic sentences) science contains only the empirical sentences of the factual sciences. Some of our opponents have seized on this and really touched a sensible spot in our overall view; they have objected that if a sentence is senseless unless it belongs to either mathematics or the factual sciences, then all the sentences in our own works are also senseless! (1934c/1987, p. 48)

In the *Tractatus*, Wittgenstein had already faced a similar question about the meaning of his own propositions and he had not drawn back from a radical solution. After asserting the meaninglessness of many an assertion of traditional philosophy, he passed a famous judgement on his own propositions in the penultimate aphorism of his book:

> My propositions serve as elucidations in the following way: anyone who understands me eventually recognizes them as nonsensical, when he has used them – as steps – to climb up beyond them. (He must, so to speak, throw away the ladder after he has climbed up it.) (1921/1961, 6.54)

Although Carnap acknowledges his debts to Wittgenstein regarding the issues dealt with in the fifth part of the book entitled 'Philosophy and Syntax', his position is in outright contradiction with Wittgenstein's on two main points underlined in §73. First, Carnap rejects Wittgenstein's idea that the syntax of language is not expressible because 'propositions cannot represent logical form' (Wittgenstein 1921/1961, 4.121): 'In opposition to this view, our construction of syntax has shown that it can be correctly formulated and that syntactical sentences do exist' (LSL, p. 282). Second, Carnap does not agree with Wittgenstein's conception of philosophy as an activity consisting of nonsensical elucidations, whatever one might understand by such activity. After recasting the propositions of philosophy in the framework of the logic of science, he means to show that 'the logic of science is syntax' and that 'the logic of science can be formulated, and formulated not in senseless if practically indispensable, pseudo-sentences, but in perfectly correct sentences' (LSL, p. 283). Indeed, one of the main motivations for Carnap's overall project in LSL is to show – against Wittgenstein – the possibility of meaningful, syntactical statements about the logical form of language.[9] In §73, after discussing at length Wittgenstein's views, Carnap also briefly mentions the solution Schlick had given to the same issue in 'The Turning Point of Philosophy' (1930). Although his views of the matter obviously contradict Schlick's position, Carnap refrains from an explicit rejection of the solution his friend had proposed a few years before (LSL, p. 284).

Carnap's solution to the aforementioned quandary lies in the method of logical syntax: 'there are no special sentences of the logic of science (or philosophy). The sentences of the logic of science are formulated as syntactical sentences about the language of science' (LSL, p. 284). If this is so, however, what should we think about all the philosophical sentences which are obviously not formulated as syntactical sentences about the language of science? As Carnap acknowledges,

> there are many sentences and questions of the logic of science which in their usual formulation appear to deal with things entirely different from linguistic structures, such as numbers, properties of numbers, mathematical functions, space and time, the causal relation between two processes, the relation between things and sense experiences, the relation between a 'mental process' and the simultaneous brain process, certain micro-processes (e.g. inside an atom) and their knowability and indeterminacy, the possibility or impossibility of some states or others, the necessary or accidental character of certain processes, and the like. (1934c/1987, p. 53)

Although these sentences, Carnap argues, seem to be object-sentences, this is only because they are formulated in a misleading way. Their translation into a more proper mode of speech would make it clear that they really have a logical character:

> Closer observation shows that such sentences only seem to refer to extralinguistic objects: they can be translated into sentences that simply talk about the formal properties of linguistic structures, i.e., into syntactic sentences. (ibid.)

Carnap terms 'pseudo-object sentences' those philosophical sentences which are logical sentences in the guise of object-sentences and which thus form an 'intermediate field' between the real object-sentences and the syntactical sentences (LSL, p. 284).

In LSL, Carnap himself frequently uses the informal and inexact mode of speech he terms 'material' (*inhaltlich*) and we too have used it in the foregoing paragraphs because this is the usual mode of speech and a convenient way to clarify Carnap's analysis of philosophy. No complete elimination of the material mode of speech is advocated:

> *It is not by any means suggested that the material mode of speech should be entirely eliminated.* For since it is established in general use, and is

> thus more readily understood, and is, moreover, often shorter and more obvious than the formal mode, *its use is frequently expedient.* (LSL, p. 312, Carnap's emphasis)

However, the material mode of speech should be *eliminable* whenever logical clarification requires it. Otherwise, Carnap's critique as implemented in LSL would be self-defeating and of very limited value. Carnap regards the translatability into the formal mode of speech as a criterion of admissibility for philosophical sentences:

> *Translatability into the formal mode of speech constitutes the touchstone for all philosophical sentences,* or, more generally, for all sentences which do not belong to the language of any one of the empirical sciences. (LSL, p. 313, Carnap's emphasis)

Again, we see here that Carnap's analysis of philosophical sentences is not purely critical: he also aims at stating precise conditions for them to be legitimate and at showing how to reconstruct at least some of them in the framework of the logic of science. For such a reconstruction, a syntactical definition of some basic concepts is indispensable. Without it, we would not be able to reformulate in a formal and unambiguous way this part of Carnap's analysis which is expressed in the material mode of speech. The key concepts here are 'quasi-syntactical sentence' and 'material mode of speech' which are given a purely syntactical definition in §63 and §64, respectively.

The formal definition of 'quasi-syntactical sentence' is meant to capture an idea that we first describe in an informal and inexact way.[10] Suppose a property $E_1$ of *non-linguistic objects* is related to a syntactical property $E_2$ of *linguistic expressions* in such a way that $E_1$ holds of an object c if and only if $E_2$ holds of a *name* of c (for any object to which $E_1$ may apply). $E_2$ is called the syntactical property correlated to $E_1$, and a sentence $\mathfrak{S}_1$ ascribing $E_1$ to c is a quasi-syntactical sentence.[11] Because $\mathfrak{S}_1$ is *equivalent* to a sentence ascribing $E_2$ to the name of c, the former is said to be translatable into the latter and is regarded as a syntactical sentence in the guise of an object-sentence.

In order to be a little bit more precise, let's take a particular case and suppose 'P' is a one-place predicate of language $S_1$. 'P' is a *quasi-syntactical predicate* if there exist a language $S_2$ and a one-place predicate 'Q' of $S_2$ such that the following conditions are met: $S_2$ contains both $S_1$ and a syntax-language of $S_1$, and for any argument 'a' of 'P', 'P(a)' is equipollent to 'Q("a")', where '"a"' is a name of 'a' in the syntax-language of $S_1$.[12 13] In this case, 'P(a)' (let's call it $\mathfrak{S}_1$) is said to be a *quasi-syntactical sentence* and 'Q("a")' (let's call it $\mathfrak{S}_2$) is said to be a syntactical correlate of $\mathfrak{S}_1$.[14] It should be noted that 'Q' may be either a logical or a descriptive predicate, so that

$\mathfrak{S}_2$ may be either a sentence of pure syntax or a sentence of descriptive syntax.[15] For a sentence $\mathfrak{S}_1$ of some language $S_1$ to be quasi-syntactical, $\mathfrak{S}_1$ must be equipollent (in some language $S_2$ of which $S_1$ is a sub-language) to some syntactical sentence $\mathfrak{S}_2$.

Now suppose the expression 'P("a")' is also a well-formed formula (i.e. the predicate 'P' may apply to a *name* of 'a' as well as to the term 'a' itself), then $\mathfrak{S}_1$ is a sentence of the *autonymous* mode of speech (in this mode of speech, the expression 'a' is used as its own designation). On the other hand, if the expression 'P("a")' is *not* a sentence (i.e. it is an ill-formed expression), then $\mathfrak{S}_1$ belongs to the *material mode of speech* and in this case, the translation of $\mathfrak{S}_1$ into the formal mode of speech is $\mathfrak{S}_2$. It is to be noted that the formal definition of 'material mode of speech' in §64 depends on the previous formal definition of 'quasi-syntactical sentence' in §63, not vice versa as some expositions of Carnap's argument erroneously assume.

In translating $\mathfrak{S}_1$ into $\mathfrak{S}_2$, we make clear that $\mathfrak{S}_1$ has the same content as a *syntactical sentence*. Carnap argues that in the material mode of speech, whereas 'P' seems to designate a property of *objects*, the fact that it is correlated (in the foregoing sense) to the syntactical predicate 'Q' clearly shows that it actually designates a property of *linguistic* expressions. The quasi-syntactical sentences of the material mode of speech

> are formulated as though they refer (either partially or exclusively) to objects, while in reality they refer to syntactical forms, and, specifically, to the forms of the designations of those objects with which they appear to deal. (LSL, p. 285)

This argument, grounded on the syntactical definition of 'quasi-syntactical sentence' in §63, is the basis of Carnap's analysis of philosophy in LSL:

> The fact that, in philosophical writings – even in those which are free from metaphysics – obscurities so frequently arise, and that in philosophical discussions people so often find themselves talking at cross purposes, is in large part due to the use of the material instead of the formal mode of speech. (LSL, p. 298)

According to Carnap, the material mode of speech deceives us in two ways. First, philosophers often erroneously believe they are dealing with objects while the issue they discuss is connected with linguistic expressions:

> Pseudo-object sentences mislead us into thinking that we are dealing with extra-linguistic objects such as numbers, things, properties, experiences, states of affairs, space, time, and so on; and the fact that, in reality, it is

> a case of language and its connections (such as numerical expressions, thing designations, spatial co-ordinates, etc.) is disguised from us by the material mode of speech. This fact only becomes clear by translation into the formal mode of speech [...]. (LSL, p. 299)

Second, the translation of philosophical sentences into the formal mode of speech makes it clear that the absolute concepts philosophers often think they are using actually are relative to some language. This is especially confusing when philosophical theses which should be interpreted as *suggestions* for the adoption of one language rather than another are mistaken for *assertions* about the nature of things. One of Carnap's examples is the philosophical controversy between the positivist who maintains that 'a thing is a complex of sense-data' and the realist who asserts that 'a thing is a complex of atoms' (LSL, p. 301). According to Carnap, the translation of these two sentences into the formal mode of speech makes it perfectly clear that their actual object is language and that they can be interpreted as suggestions regarding the best way to build the language of science. Carnap concludes that '*the controversy between positivism and realism is an idle dispute about pseudo-theses which owes its origin entirely to the use of the material mode of speech*' (ibid.).

The examples of sentences in the material mode of speech that Carnap quotes are taken from very varied domains of philosophy (LSL, §79). Two particular cases, however, are given a special emphasis: sentences about meaning (LSL, §75) and universal words such as 'thing', 'object', 'property', 'relation', 'fact', 'number'... (LSL, §§76–7).

A definition of universal words is given on pp. 292–3 of LSL. For the sake of simplicity, let's consider the word 'number' as a particular case. Not all expressions are legitimate arguments of the sentential function '...is a number'. For example, 'the moon' is not a legitimate argument because 'the moon is a number' is nonsense. The point is that for any legitimate argument, the resulting sentence is analytic. By contrast, the sentence resulting from the application of '... is an odd number' to a legitimate argument may be either analytic or contradictory. So the predicate 'number' is universal while 'odd number' is not. According to Carnap, the trouble with universal words is that they easily lead to pseudo-problems when used as quasi-syntactical predicates in the material mode of speech because in such uses, philosophers easily mistake syntactical questions for object-questions. Philosophical inquiries about the nature of numbers provide a typical example:

> Philosophers from antiquity to the present day have associated with the universal word 'number' certain pseudo-problems which have led to the most abstruse inquiries and controversies. It has been asked, for example, whether numbers are real or ideal objects, whether they are extra-mental

> or only exist in the mind, whether they are the creation of thought or independent of it, whether they are potential or actual, whether real or fictitious. (LSL, p. 310)

All these questions – that Carnap regards as pseudo-questions – are eliminated by translation into the formal mode of speech:

> All pseudo-questions of this kind disappear if the formal instead of the material mode of speech is used, that is, if in the formulation of questions, instead of universal words (such as 'number', 'space', 'universal'), we employ the corresponding syntactical words ('numerical expression', 'space-co-ordinate', 'predicate', etc.). (LSL, p. 311)

It must be acknowledged that this kind of criticism had a limited impact, to say the least, on discussions about numbers among philosophers. In a lecture given in 1937, Quine distinguished two purposes of the nominalist stance in philosophy: the first is to avoid metaphysical issues regarding the connections between universals and particulars; the second is to achieve a reduction of any sentence to sentences about 'tangible things' so as to avoid 'empty theorizing'.[16] The dismissal of any philosophical discussion about universals such as numbers is Carnap's answer to the first issue. However, as Quine remarked, a consequence of this dismissal is that Carnap has very little to say about the second issue: the reduction of scientific statements to statements about concrete things (Mancosu 2008a, pp. 28–9). Carnap would probably respond that the reason why he has 'little to say' about this question is that, taken literally, it is a pseudo-question. Indeed, its translation into the formal mode of speech would make it clear that it really is a linguistic issue: the reduction depends on the choice of a language. Like Tarski and Quine, however, many philosophers were not at all convinced by Carnap's arguments that all metaphysical discussions about numbers were just idle philosophizing breeding talks at cross purposes. As a matter of fact, such discussions are still flourishing today.[17]

Another special case of sentences of the material mode of speech is given emphasis in §75: sentences about meaning, content, designation, and sense. In this paragraph, Carnap explains through examples how to get rid of expressions such as 'treats of', 'means', 'signifies', 'is about', or 'designates' by translating into the formal mode of speech the sentences in which they occur. The elimination of sentences which we may regard as typical of a *semantic* mode of speech is hardly surprising in a book whose leitmotif is 'logic is syntax'. The trouble is that the translations Carnap suggests are not all equally convincing. In 1936 already, in one of the first extensive examinations of Carnap's rejection of semantic concepts in LSL, Maria Kokoszyńska levelled doubts at Carnap's translation of 'yesterday's lecture was about Babylon' into the formal mode of speech as 'the word "Babylon" occurred in yesterday's lecture'. Obviously, contrary to what Carnap assumes, the former sentence can be perfectly true even if the word 'Babylon' did not occur a single time in 'yesterday's lecture' (Kokoszyńska 1936, p. 160).

Kokoszyńska raises another objection, this time against Carnap's argument that quasi-syntactical sentences of the material mode of speech *seem* to refer to objects 'while *in reality* [*in Wirklichkeit*] they refer to syntactical forms' (LSL, p. 285, my emphasis).[18] Recall that a sentence $\mathfrak{S}_1$ of language $S_1$ is quasi-syntactical only if there is a language $S_2$ of which $S_1$ is a sublanguage and a syntactical sentence $\mathfrak{S}_2$ of $S_2$ equipollent to $\mathfrak{S}_1$. Now why should we conclude with Carnap that $\mathfrak{S}_1$ only *seems* to refer to objects and *in reality* refers to syntactical forms? Since 'equipollent in $S_2$' is a symmetrical relation, why couldn't we conclude the exact opposite? Kokoszyńska further argues that it is not at all easy to give a clear meaning to expressions such as 'seems to be about' or 'in reality refers to' anyway (1936, pp. 163–4).

As a typical example of a sentence about meaning in the material mode of speech, Carnap gives 'this letter *is about* the son of Mr. Miller', which he proposes to translate into the formal mode of speech as 'in this letter a sentence $\mathfrak{Pr}(\mathfrak{A}_1)$ occurs in which $\mathfrak{A}_1$ is the description "the son of Mr. Miller"' (LSL, p. 290).[19] Now, assuming that Mr. Miller actually has no son, Carnap shows how to infer a false sentence from the former sentence by using the ordinary rules of logic and he then argues in favour of the formal mode of speech that the same falsity cannot be derived from the latter sentence. A direct consequence of Carnap's argument, however, is that the two sentences in question are not equipollent (because they do not have the same consequences), so that the latter *cannot* be regarded as the syntactical correlate of the former in the first place.

This particular example raises a more general issue about Carnap's critique of philosophy in LSL. On the one hand, this critique is based on syntactically defined concepts such as 'quasi-syntactical sentence' and 'material mode of speech' which receive a formal definition in Part IV (about 'general syntax'). Indeed, without such a formal basis, the distinction between the material and the formal modes of speech would be inexact and of limited value, Carnap's critique of misleading sentences in the material mode of speech would apply to itself and thus be self-defeating, and the syntactical method would offer no way out of the quandary that the logical empiricists had to face, regarding their own philosophical sentences (see above, p. 190). So the tools of Carnap's critique here are formal tools defined in the context of 'general syntax'. And so they had to be. On the other hand, the philosophical sentences that Carnap analyses throughout Part V and that he interprets as misleading formulations of the material mode of speech usually belong to a natural word-language which is not known to us as formally defined by formation and transformation rules. So the general issue is: how can a formal tool designed for sentences of formal languages be applied to philosophical sentences formulated in natural word-languages?

Before undertaking the exposition of general syntax in Part IV, Carnap gives the following warning:

> As opposed both to the symbolic languages of logistics and to the strictly scientific languages, the common word-languages contain also sentences whose logical character (for example, logical validity or being the logical consequence of another particular sentence, etc.) depends not only upon their syntactical structure but also upon extra-syntactical circumstances. [...] In what follows, we shall deal only with languages which contain *no expression dependent upon extra-linguistic factors*. (LSL, p. 168, Carnap's emphasis)

Because the concepts 'quasi-syntactical sentence' and 'material mode of speech' are defined in the context of general syntax, they are relative to any language, but in this context, this means to any set of formal and transformation rules. A concept such as 'equipollent in language S′', on which the definition of the two concepts just mentioned depends, has no precise meaning as long as no formal rules for S have been provided.

How shall we decide whether some given philosophical sentence $\mathfrak{S}_1$ formulated in a word-language such as English is quasi-syntactical? According to Carnap's definition, $\mathfrak{S}_1$ is quasi-syntactical only if there is a language $S_2$ containing both English and a syntax-language of English as sublanguages.[20] So far so good: English itself contains both English and a syntax-language of itself as sublanguages. A further condition, however, is that $S_2$ contain a syntactical sentence $\mathfrak{S}_2$ equipollent to $\mathfrak{S}_1$. This is the point where we get into trouble because English is not known as a language defined by formation and transformation rules and 'equipollent in English' does not have any precise meaning as long as English has not been defined by such rules. Therefore, we actually have no way to decide whether $\mathfrak{S}_1$ is quasi-syntactical or whether $\mathfrak{S}_1$ is formulated in the material mode of speech. In order to make tentative decisions on these points, we can rely on informal and inexact characterizations of these concepts, and this is what Carnap actually does as far as sentences formulated in a word-language are concerned. However, the whole syntactical machinery of Part IV is then left aside, unused, the critique loses its grip, and objections similar to the one we formulated against Carnap's analysis of 'this letter *is about* the son of Mr. Miller' can be raised against any other examples.

Carnap is perfectly aware of the situation since in the very paragraph in which 'material mode of speech' is formally defined, he already notes that

> the decision that certain sentences are quasi-syntactical (not genuinely syntactical) can be made with the same degree of exactitude with which the language in question is itself constructed. (LSL, p. 240)

If we consider that the word-languages in which the philosophical sentences are usually formulated are constructed with a very low degree of logical exactitude, Carnap's remark seems to entail that we have no solid ground for applying the principles of his analysis of philosophy in LSL to actual philosophical sentences. Carnap himself admits that there is a real issue:

> The examples of sentences which come later, especially those of the logic of science, belong almost entirely to the word-language; in consequence, they are themselves not formulated sufficiently exactly to make possible the application to them of exact concepts. (LSL, p. 287)

The fact that Carnap nevertheless provides no further justification for his applying formally defined concepts to philosophical sentences formulated in ordinary word-languages weakens his analysis of philosophy as formulated in Part V.

A similar objection can be raised against Carnap's analysis of universal words, a central part of his critique of traditional philosophy. Here again, the difficulty of applying formal concepts to the syntax of word-languages is made explicit by Carnap himself:

> Since the rules of syntax of the word-language are not exactly established, and since linguistic usage varies considerably on just this point of the generic classification of words, our examples of universal words must always be given with the reservation that they are valid only for one particular use of language. (LSL, p. 293)

In his analysis of '5 is a number', Carnap takes for granted that 'number' is a universal word in the language in which the sentence is formulated although this is questionable. We do know languages in which the concept 'number' is *not* a universal word (on p. 293, Carnap cites Russell's language as an example in which this is the case). Again, a decision on whether 'number' is a universal word in English would require fixing formation and transformation rules for this language.[21]

In his analysis of Carnap's idea of a translation from the material into the formal mode of speech, André Carus notes that 'the language *in* which the translation is done (stated) is also the language ($L_2$) *into* which the translation is to be made (from $L_1$)' (Carus 2007, p. 259) and he proposes the following interpretation:

> Wittgensteinian scruples about the impossibility of stepping outside the language still prevented [Carnap] from considering a meta-language for statement and discussion of the translation that was distinct from the target language for rational reconstruction (or explication). (ibid.)

André Carus interprets this translation as a first and still inadequate attempt to formulate what will become the ideal of 'explication' in Carnap's later works.[22] It may be added that the translation is internal in still another sense since it is defined in such a way that $L_1$ is a sublanguage of $L_2$ so that the translation actually takes place inside $L_2$. This may be contrasted with Tarski's semantic work of the same period. In 'The Concept of Truth in Formalized Languages' (1933), Tarski distinguishes an object-language L, a metalanguage ML essentially richer than L and containing a name and a translation of any expression of L, and a meta-metalanguage in which the conditions of adequacy for the definition (in ML) of 'true in L' are formulated. Soon after the publication of LSL, Carnap heard about Tarski's definition of truth and enlarged the syntactical approach to include a semantic one. In this new phase of his thinking, he still maintained that philosophical sentences often mislead us into believing we are dealing with non-linguistic objects while the object of our discourse is language, although the basis for this diagnosis had changed:

> Many sentences in philosophy are such that, in their customary formulation, they seem to deal not with language but merely with certain features of things or events or nature in general, while a closer analysis shows that they are translatable into sentences of L-semantics. Sentences of this kind might be called *quasi-logical* or cryptological. (1942, p. 245)

One of the striking differences with LSL is that no exact and technical definition of what is to count as a *translation* from the material into some formal mode of speech is provided anymore. In this context, logical syntax is but one part of semiotics, a general study of language which also comprises semantics and pragmatics. Carnap's retrospective comments about Part V of LSL give evidence of the evolution of his views:

> The *chief thesis* of Part V, if split up into two components, was this:
>
> a. '(Theoretical) *philosophy* is the logic of science.'
> b. 'Logic of science is the syntax of the language of science.'
>
> (a) remains valid. [...] Thesis (b), however, needs modification by adding semantics to syntax. Thus *the whole thesis is changed to the following: the task of philosophy is semiotical analysis*; [...]. (1942, p. 250)

Although Carnap soon realized his analysis of philosophy needed a more general basis than the syntactical one he had elaborated in LSL, he by no means gave up his idea that the problems of philosophy are at bottom problems of language:

> It has turned out to be very fruitful to look at the problems of theoretical philosophy from the point of view of semiotic, i.e. to try to understand

> them as problems which have to do with signs and language in one way or another. (1942, p. 245)

In later works, the basis of his analysis would become his famous distinction between internal and external questions.[23]

Still later, in his response to Charles Morris in the Schilpp volume, Carnap made the following remarks about the evolution of his thoughts regarding philosophy:

> In earlier periods, I sometimes made attempts to give an explication of the term 'philosophy'. [...] Yet actually none of my explications seemed fully satisfactory to me even when I proposed them; and I did not like the explications proposed by others any better. Finally, I gave up the search. I agree with Morris that it is unwise to attempt such an explication because each of them is more or less artificial. It seems better to leave the term 'philosophy' without any sharp boundary lines, and merely to propose the inclusion or the exclusion of certain kinds of problems. (1963b, p. 862)[24]

In (1950a), Carnap gave a final account of the kind problems he meant to include in his own research that we may still call 'philosophical questions'.

The principle of tolerance is, of course, another key point for understanding the evolution of Carnap's thought both before and after the publication of LSL, for both Carnap's critique of metaphysics and his analysis of 'theoretical philosophy'. These particular aspects of Carnap's thought are dealt with at length in other parts of this book.[25] Steve Awodey and André Carus analyse the method of logical syntax into several theses and show which part of the method is compatible with the principle of tolerance and which part is not (see above, pp. 100–1). They also note that the principle of tolerance is almost totally absent from Part V, where Carnap gives his fuller account of the impact of the syntactical method on philosophy. In his contribution, Richard Creath assesses the exact consequences for Carnap's philosophical programme of what he calls 'the gentle strength of tolerance'.[26]

## Notes

1. Such engagement is nevertheless made explicit in the first preface of the *Aufbau*, in the *Vienna Circle Manifesto*, and in his 'Intellectual Autobiography' (1963a). See Uebel (2004) and Carus (2007).
2. Note that this is different from what we call 'analytic metaphysics' nowadays. Whether Carnap would have approved of analytic metaphysics more than he did of the German idealist or phenomenological kind is another question.
3. See above, Introduction, pp. 12–17.
4. Carnap also gives examples pertaining to the foundations of mathematics; see LSL (p. 300).
5. On this distinction, see above, Introduction, pp. 7–8.

6. The linguistic character of philosophical theses formulated in the framework of the logic of science is analysed above, in the Introduction to this volume, pp. 6–8.
7. The distinction between the material and the formal modes of speech is first made in print in Carnap (1932b).
8. For a more elaborate and qualified statement of this point, see Uebel (2007a).
9. This issue had been much discussed in the Vienna Circle in 1930–1. See Awodey and Carus (2007, pp. 26–8), Carus (2007, ch. 9), and Steve Awodey and André Carus's contribution to this volume.
10. Carnap gives an informal exposition on this concept on pp. 234 and 287, and a formal definition on pp. 235–6.
11. On Carnap's use of gothic letters, see above, Introduction, p. 18 and pp. 43–5.
12. This definition depends on a previously given one-one syntactical correlation $\mathfrak{Q}_1$ (to be read '$Q_1$') between the expressions of $S_1$ and expressions of the syntax-language of $S_1$ included in $S_2$. If $\mathfrak{A}_1$ is an expression of $S_1$, $\mathfrak{Q}_1[\mathfrak{A}_1]$ is its name, or its 'syntactical designation' (LSL, p. 235). Because $S_1$ is a sublanguage of $S_2$, both $\mathfrak{A}_1$ and $\mathfrak{Q}_1[\mathfrak{A}_1]$ are expressions of $S_2$. For the sake of simplicity, I use here quotation marks to denote names.
13. Two sentences of language S are equipollent (*gehaltgleich*) in S if they have the same '(logical) content' (*logischer*) (*Gehalt*) in the following sense: each non-valid sentence which is a consequence (in S) of one of them is also a consequence (in S) of the other (LSL, pp. 42, 120, and 176).
14. The condition '$S_1$ is a sub-language of $S_2$' (LSL, p. 236) is crucial. Otherwise 'P(a)' and 'Q("a")' would not necessarily be sentences of the same language and there would not be any sense in saying they are equipollent. When Carnap says that 'P(a)' is equipollent to 'Q("a")', he means equipollent *in* $S_2$. Compare with André Carus's analysis in Carus (2007, p. 257).
15. In the examples given on pp. 234–5 of LSL, the quasi-syntactical sentences are logical sentences, but in Part V Carnap gives examples of quasi-syntactical sentences the syntactical correlate of which are sentences of descriptive syntax. See for example 'the word "Babylon" occurred in yesterday's lecture' (LSL, p. 286).
16. This unpublished lecture is quoted by Paolo Mancosu in Mancosu (2008a).
17. See Paolo Mancosu's analysis of the discussions that Carnap, Tarski, Quine, and other philosophers had about a finite language for mathematics and science in 1940–1, in Mancosu (2005).
18. This kind of remark is repeated on several occasions. See for example LSL (p. 312).
19. The expression '$\mathfrak{Pr}(\mathfrak{A}_1)$' (to be read '$Pr(A_1)$') designates a sentence of the object-language resulting from the application of a predicate-expression to an expression $\mathfrak{A}_1$ (to be read '$A_1$').
20. A formal definition of '$S_2$ contains a syntax of $S_1$' is given in §63 (LSL, p. 235).
21. The reason why 'number' has to be a universal word for the sentence '5 is a number' to be quasi-syntactical is rather subtle. See the end of §63 (LSL, pp. 236–7). On the sentence '5 is a number', see André Carus's analysis in Carus (2007, pp. 257–8).
22. This ideal makes its first appearance in Carnap (1945) and finds its most well-known exposition in Carnap (1950b). See Carus (2007, p. 256) and Beaney (2004).
23. The classical exposition of this distinction is Carnap (1950a). A fuller account of Carnap's critique of philosophy after LSL goes beyond the limits of this chapter. See Goldfarb (1997, pp. 63–5), Friedman (1999b, pp. 215ff.), Carus (2007, pp. 263ff.), and other contributions to this volume.

24. I am indebted to André Carus for drawing my attention to this quotation.
25. See the Introduction, as well as Steve Awodey and André Carus's, Jacques Bouveresse's, Richard Creath's, Michael Friedman's, and Thomas Ricketts's contributions.
26. I am grateful to the participants in the 2005 Paris colloquium about LSL for helpful discussions and to André Carus for his most valuable comments on an earlier draft of this chapter.

# 8

# The Gentle Strength of Tolerance: *The Logical Syntax of Language* and Carnap's Philosophical Programme

*Richard Creath*

Before Rudolf Carnap wrote *The Logical Syntax of Language* he was an important philosopher, a very important one. After he wrote it, he was a great one. And tolerance is his great idea; it is at the very centre of his philosophic programme from this moment on. With most philosophers a programme is a wellspring of ideas that occasionally crystallizes into a book. In this case, the order is reversed: the book produced the idea and the idea generated the programme.

To see how this might be so I will look first at how the book developed. Even though parts of the finished book do not represent Carnap's mature ideas or programme and even disguise them, the historical development throws light on both the book and its outcome. Second, I will examine tolerance itself to see what opportunities it offers for empiricism and for so-called verificationism. Along the way it will be necessary to distinguish two different senses of 'tolerance'. Third and finally, I want to explore the new programme for philosophy that emerged from tolerance in *Logical Syntax*.

## 1 Empiricist origins

Thanks to various archives, Carnap's own autobiography, and much recent scholarly work, most notably a fine new paper by Awodey and Carus (2007), we know much about how *Logical Syntax* came to be – not everything, but we know a lot. We know for example that after Carnap moved to Vienna he and his colleagues studied Wittgenstein's *Tractatus* (Wittgenstein 1921) with great care. They were particularly attracted to his 'no content' account of logic because it seemed to make empiricism, to which they were attracted in any case, a viable account of all our knowledge. Our science seems to require over and above its clearly empirical part at least logic and mathematics. Logicism may assure us that mathematics reduces to logic, but that still leaves logic to be accounted for. If logic has a clear content, perhaps as the most general theory of objects, as some might assume, then how do we know that content? If the only answer is Platonic intuition or some other form of metaphysical insight, then empiricism is in real trouble. So Wittgenstein's idea that logic is without content was attractive.

It was attractive in another way. There had been a long and vocal tradition in the scientific community of impatience with philosophers who speculate about matters that can never be resolved and ignore urgent empirical issues that can. These philosophers they derided as metaphysicians. On the whole, Carnap's sympathies lay with the scientific community and especially with the task of getting on with the work that could be done rather than engaging in fruitless wrangles. Here the *Tractatus* was suggestive. According to Wittgenstein, it is the atomic sentences that most directly picture the world. The molecular sentences will be completely settled by the atomic ones, and the quantified sentences will be problematic depending on how they are understood. So it seemed that the only intelligible sentences were completely settled by the atomic ones. And the truly intelligible sentences are to be found only in the natural sciences. Read in a certain way, Wittgenstein's position amounts to an extremely strong verificationism. I am not inclined to see this reading as right, but it is not hard to see why the Viennese were tempted.

Both Wittgenstein (*Tractatus*, 6.53) and the Viennese rejected metaphysics as unintelligible. Both held that it was unintelligible precisely in so far as it purports to provide substantive truths above or beyond those of the natural sciences. Both rejected it on the grounds that the supposed metaphysical claims failed to have the right logical relation to certain atomic sentences taken as basic. For Carnap and company these basic sentences were protocol or observation sentences. What Wittgenstein meant by 'atomic sentences' was far less clear. Since Wittgenstein seemed to reach a similar conclusion via a similar looking argument, the Viennese were inclined to read Wittgenstein as meaning by his atomic sentences what they meant by their observation sentences. Given this interpretation, Wittgenstein's position would indeed amount to a robust verificationism. The presumption of philosophers to go beyond science to a deeper truth was not just a futile waste of time; it was not even false; it was utterly unintelligible. Perhaps this was a tad stronger than we need, but watching the metaphysicians squirm was undeniably delightful.

There are two things to note here: First, I am emphatically not endorsing either this reading of the *Tractatus* or the *Tractatus* itself. In fact, I reject Carnap's reading of that book, but it does not make a bit of difference to the issue at hand whether Carnap was right in his understanding of Wittgenstein. My concern is entirely with what the Viennese thought they saw there. In their defence I will say only that the interpretation of the *Tractatus* is notoriously difficult, and Wittgenstein himself is not particularly helpful on the question of atomic sentences. Second, when I speak of metaphysics here I do not mean all or even most of what *now* goes under that term. I mean rather the idea that philosophers have their own special mode of access to a world that is deeper or 'behind' the world described by empirical scientists.

In any case the good news for the empiricists and bad news for the metaphysicians that the Viennese saw in the *Tractatus* came with a price. There it was attached to the idea that one could not fully and intelligibly talk about logic. Logic is, to say the least, rather general, and if it could be reduced to a collection of observation sentences, it would never have been a problem for empiricism in the first place. So officially, logic and even the *Tractatus* itself is unintelligible. If that book contains a verifiability principle, then that principle is not itself intelligible either. Wittgenstein is not shy about admitting the irony of his position, though in his defence perhaps it could be urged that even if he cannot *say* anything about logical structure, he can try to *show* it. In this he is undoubtedly more virtuous than those who merely misuse language. Wittgenstein, it seems, had more of a taste for irony than did the Viennese empiricists; they found it uncomfortable.

We come then to the second stage in the creation of *Logical Syntax*, the famous sleepless night of January 1931 (see Carnap 1963a, pp. 53f.). In essence what Carnap decided that fateful night was that the metamathematical point of view, exemplified by Hilbert, Gödel, and Tarski, could be adopted to talk about logical form. This could apparently still be combined with the idea that logic does not tell us anything about the world. By deploying an analytic-synthetic-contradictory distinction, one could coherently argue that only the synthetic sentences have content while the analytic and contradictory sentences are settled by the rules of the system, not the world. On this approach, logical sentences are genuine, intelligible, sayable sentences. Once truth is readmitted officially, they will be counted as true. In the interim, 'analytic' can function as a truth predicate within the logical domain. As of 1931 tolerance was still not part of the story.

On this new syntactical approach there were a great many new problems to be worked out such as whether and to what extent claims about the logical form of sentences in a given language could be stated in that very language. Fortunately, that particular problem need not detain us. Another new problem, however, cannot be sidestepped. Now that we can talk about logical form, we can talk about alternative logical forms and alternative logical systems. So the question arises or seems to arise as to which is correct, and means for answering such questions do not arise along with it. There had been no corresponding problem for Russell for in his non-empiricist moods he could always appeal to direct acquaintance with universals to choose among systems, should alternative systems be recognized. And for Wittgenstein in the *Tractatus* there simply are no alternative systems. This does not reflect that logic is unsayable but is rather a deeper result of the way that language pictures the facts. The world is the totality of facts. There is only one world; so there is only one totality of facts. Since linguistic form reflects the form of those facts, there is only one form of language. The question of which form is correct does not arise. Once Carnap adopts the metalinguistic point of view, derived from the metamathematicians, he can no longer avoid the problem à la Wittgenstein or solve it à la Russell. My own suspicion is that when he speaks in the 'Foreword' about the dangers of thinking that one or another logical system is correct (1934d/1937, pp. xiv–xvi/ pp. iv–vi), he is at least also thinking about himself in this period between the sleepless night of January 1931 and his acceptance of the principle of tolerance.

When we spoke earlier about the *Tractatus* we noted that from the perspective of the Viennese empiricists it seemed to embody a very strict verificationism that was (again on the Viennese view) unfortunately unsayable. With the move to the syntactical point of view one could state a verificationist principle, though one would have to do so carefully: intelligible claims are either analytic (or contradictory) or verifiable. The change also brings room to manoeuvre on verifiability; it need not be as strong or restrictive as it seemed in the *Tractatus*. But there is still a worry as to whether the principle is self-undercutting in another way. Is the principle itself analytic? Well, not in any of the formal systems standardly discussed. Is it verifiable, even in some weak sense? That would seem to require an independent criterion of verifiability that we lack. And if we had one, would the principle even turn out to be true? In the absence of tolerance I do not know how to respond to such questions. But I cannot argue that such considerations led Carnap to tolerance. I am not aware that there is documentary evidence that shows precisely what led Carnap to tolerance. But we do know roughly when he got there, namely 1932, probably after late September.[1]

While the principle of tolerance, as such, was new, the turn of mind that led Carnap to play a harmonizing role – among philosophers he respected – had been present for many years, at least since his dissertation and probably before that. Moreover, in his scientific work, including especially his thinking about geometry, Carnap was already disposed toward some sort of pluralism, even toward conventionalism. One might describe this world as Euclidean or not, depending on the metric chosen. And it *is* a choice. This pluralism in geometry and even the idea that one might choose among the alternatives on such pragmatic grounds as simplicity is encouraged by the work of Hilbert, Poincaré, Duhem, and Einstein. But the same opportunity for choice did not extend to logic. No doubt this was a residual influence of the logicist tradition. Carnap's teacher, Frege, had had a rather nasty fight with Hilbert,[2] and the young Russell had fought with Poincaré (Coffa 1991, pp. 129–34). In logic, Carnap's presuppositions were against the pluralism of Hilbert and Poincaré. But now – slowly – he was beginning to see the light.

Let us turn then to look directly at the principle of tolerance in order to discover its surprising and gentle strength. We will find that a principle that has the precedent just discussed in Carnap's philosophy of science has some of its application there as well.

## 2 The gentle strength of tolerance

Tolerance is the third, final, and most important stage in the creation of *Logical Syntax*. It made its dramatic appearance in the context of the protocol sentence debate (Carnap 1932c). One side in the debate at hand can be thought of as holding that observation sentences (protocol sentences) are about phenomena or the given and that these sentences are certain or unrevisable. The other side can be thought of as holding that these sentences are about physical objects and events and are as revisable as any other hypotheses. Given the natural reading of the foregoing positions the conflict between the sides here is quite direct. But Carnap does not come down on one side or the other, but against the contradiction. In 'On Protocol Sentences' he remarks:

> My opinion here is that this is a question, not of two mutually inconsistent views, but rather of two different methods for structuring the language of science, both of which are possible and legitimate. (1932c/1987, pp. 215/457)

And then shortly later he says:

> [...] I now think that the different answers do not contradict each other. They are to be understood as suggestions for postulates; the task consists in investigating the consequences of these various possible postulations and testing their practical utility. (1932c/1987, pp. 216/458)

That Carnap thinks the two sides in the debate do not contradict one another is overdetermined here. First, the two sides are proposals and hence not assertions and hence not even candidates for contradictories of one another. Second, even though there are associated assertions, these do not contradict each other either. This is because the proposals are methods of constructing the language of science. Hence, the two associated assertions would be in two separate languages and so not in a position to contradict each other. While there are many languages, and possibly nonequivalent ones, there is no question of the correctness of the languages. We are left with only the issue of the pragmatic usefulness of a given language.

Curiously, by the time that tolerance made its appearance, much of *Logical Syntax* had already been written, including most of Part V, a fact that makes it very difficult for the reader. I will argue here that the principle of tolerance is a powerful new tool for defending empiricism and the verifiability principle, but on the surface it must seem to be just the opposite. Empiricism and verifiability had once had real bite to them. Now it seems that Carnap is prepared, in the name of tolerance, to weaken strongly held philosophical positions into mere proposals and to demote them to a status of one among many, none better than the rest. We can resist this characterization, but I can understand why it might seem so.

This appearance of tolerance as a form of weakness is encouraged, at least in English, where standard definitions of 'tolerate' render it as 'endure without hindrance or prohibition'. It is this meaning of 'tolerance' that Quine trades on when he writes to Carnap in 1938: 'As I told Hempel, I fear your principle of tolerance may finally lead you even to tolerate Hitler.'[3] Well, Quine was being unfair, as he was apt to be when discussing intensional languages. This is most certainly *not* the sense of tolerance that Carnap had in mind in his principle.

Even in the new regime it is plain that Carnap has no intention whatsoever of enduring metaphysics without hindrance. And while he is soon to call empiricism a convention[4] and to treat verifiability as a proposal, it is also plain that he still intends to campaign quite vigorously for both. What has changed is the form that that campaign would take. 'Tolerance' in Carnap's sense involves treating alternative philosophical positions *if they were to be made clear and precise,* in the first instance as proposals for structuring the language of science. If a proposal is adopted, then there will be features in the language that reflect features of the proposal. In the language of *Logical Syntax* these features will be expressed in sentences that are L-valid or P-valid. Later, that distinction drops out, and he uses just 'L-true' or 'analytic' instead. Within one language one can speak of correctness (or something relevantly like it), but that does not compare philosophical positions. That kind of comparison, which is in effect from one language to another, can be given in pragmatic terms only. Just because the comparison was to be in pragmatic terms, however, does not mean that it could not be extremely pointed, had Carnap so chosen.

I want now to illustrate this point by showing that the defence of empiricism and verifiability (they are of course related) can be bolstered by the gentle strength of tolerance. I shall proceed as Carnap would not have: very briefly and in broad strokes. My version of empiricism here is broad and generous too, saying little more than that our information about the world comes solely from ordinary observation. I mean to include the empiricism of Locke and Hume as well as that of Quine and Neurath. But it has some bite, for I mean to exclude Gödel's account of mathematical knowledge, Russell's theory of direct acquaintance, Plato's doctrine of recollection, and as much of American evangelical talk radio as I possibly can. Actually I mean this last seriously because it is merely a public version of the 'wearisome controversies' for which metaphysics in the special sense here at issue is an academic version. And it too purports to get beyond or behind the world described by science.

If empiricism is dogmatically assumed it invites dogmatic rejection. If empiricism is said to be a result of empirical science, it will be taken to beg the question. So, Carnap says, this is what I propose clearly and explicitly. If you have an alternative, let us see it with equal clarity and explicitness. Only then can we evaluate their relative merits. Not only will I allow you to make your own proposal in your own terms, as long as you do it clearly, says Carnap, but I will forebear calling your proposal false and mine true. Of course, you must do the same. Once the rules of the languages are laid out they can be explored with engineering exactitude to see what the pragmatic consequences of using them would be for science or, if you like, for the greater good of mankind. What could be fairer than that?

This puts Carnap in an extremely strong rhetorical position. But it is better than that. Even if we discount the extreme improbability of most metaphysicians rising to the challenge of great clarity, Carnap is convinced that the historical evidence is overwhelmingly on his side in the proposed pragmatic investigation. Metaphysical wrangles never get anywhere. They may seem to as one side or another becomes fashionable, but in the long run they are just, to use Carnap's word, 'wearisome'. Engineering and logic and empirical science by contrast have gotten somewhere. So even if Carnap cannot call the metaphysical point of view false, he can argue that it is unwise, foolish, misguided, imprudent, and utterly futile. Soon the metaphysician will plead for no more of this toleration and beg to be dismissed once again as false. My point in all this is that in moving to the principle of tolerance Carnap is not in fact weakening his position with respect to empiricism. He does not have to sit idly by and endure its rejection. He can, and does, campaign vigorously for us all to embrace some form of empiricism. All that changes is the form of the campaign.

I have spoken twice already of verifiability, once in noting that the *Tractatus* was taken to embody a very strong version that would be unsayable. With the advent of the syntactical or metalogical point of view some verifiability principle is statable, but actually stating one in a way that it is not self-undercutting is a delicate matter. Suppose we say that to be intelligible a claim must be either analytic (or contradictory) or verifiable in some way. That last might mean no more than that it is confirmable, but we need not worry about that here. We do need to worry, however, about the status of the verifiability principle itself. It has generally seemed not to be analytic or verifiable, and if it turned out to be contradictory that would hardly come as good news to its promoters. There are still philosophers who think that the verifiability principle is permanently doomed for precisely these reasons.

I am not among them, and I think that the principle of tolerance provides a perfectly viable way out and that Carnap even hinted at this in the brief passages I quoted earlier from 'On Protocol Sentences'. I think that what we are supposed to do is to think of the verifiability principle first as a proposal that we adopt a certain language, call it L. If the proposal is adopted, then it will be true that all sentences of L will be analytic, contradictory, or properly confirmable. This truth about L will be an analytic claim in a properly constructed metalanguage for L, namely ML. And this analytic sentence of ML is the associated sentence of the verifiability principle itself. In this way the verifiability principle is completely and intelligibly statable and in no way self-undercutting. I have separated the object and metalanguages here for the purposes of clarifying the exposition. If they can be combined as Carnap does in Language I of *Logical Syntax*, then that is an independent matter. In any case, our ability not merely to talk about language but also to make proposals for altering it is crucial. Of course, tolerance also provides Carnap with a strategy for defending the proposal involved in the verifiability principle. The language proposed is, after all, nothing more or less than an empiricist language, and the same pragmatic defence of empiricism discussed earlier is available likewise for defending the verifiability principle.

In sum, then, Carnap only appears to be weakening his position in adopting the principle of tolerance. He is not even remotely forced to sit helplessly by unable to resist opposing alternative philosophical positions. Instead, his defence of empiricism is powerful, and he is finally in a position to state and defend a coherent version of the verifiability principle. But tolerance gives him something more. It opens up what he calls 'the boundless ocean of unlimited possibilities' (1934d/1937, p. xv/p. vi). In short, it gives him a self-conception and a direction for future research.

## 3 A positive programme

We come at last to Carnap's philosophic programme – the one he had coming *out* of *Logical Syntax*, the one that was made possible by adopting tolerance. Before I say what that new programme was, I want to contrast it with another view, call it the 'received unwisdom'. This is the extremely narrow and destructive programme for philosophy that he is often assumed to have. On the received unwisdom, philosophy is nothing more than technical logic, entirely stripped of any human interest beyond pure mathematics. Metaphysics should be overthrown. Ethics should be abolished. Even epistemology turns out to be tainted with psychology and must be discarded. What is left is thin and bleak. As we know, many people outside philosophy think that all analytic philosophy is of this sort, and many analytic philosophers still take the received unwisdom to be Carnap's actual programme. To be sure, *The Logical Syntax of Language* had plenty of technical logic in it, and one can find scattered remarks, and sometimes more than that, to bolster the received unwisdom. But attributing this view to Carnap misses the whole point.

In fact I think the idea that Carnap's programme for philosophy is narrow and destructive is almost exactly wrong. Indeed his programme is a reconception of philosophy itself, and of its methods, a reconception according to which it is possible for philosophy to make stepwise progress and according to which philosophy can play a positive, constructive, and progressive role in science and in society at large. This is the programme that tolerance opens up.

The first thing we have to recognize is that when Carnap speaks of rejecting metaphysics and other branches of philosophy like it, he is not rejecting all existence claims, truth claims, and proposals for practical action. Later, when he is equally anti-metaphysical, he would cheerfully quantify over properties and propositions, and over physical objects and numbers, that is, he would say that there are such things. But of course he would do so always *within* a framework. Early and late he is warning us against the absolute pronouncements that pretend to go behind or deeper than a proper science can go. These absolute pronouncements are the purported answers to what he would later call 'external questions' (Carnap 1950a). In *Logical Syntax* this is put in terms of rejecting the standpoint that there is a correct logic.

If philosophers were to take the opposing, un-Carnapian, stand, that is, to reject tolerance and seek correctness in logic, system, and metaphysical view, they would have little choice but to use direct metaphysical insight as a method or to hope that we were all inclined to impose the same systematic forms on the matter of experience. The historical track record here has been grim. There is no reason to think that progress is possible. Instead, we must resign ourselves to a philosophical discourse that will be little more than comparative autobiography and so-called progress little more than the random walk of fashion.

Against this, Carnap is offering a model of conceptual engineering. On this model, philosophers can devise, refine, and explore a variety of conceptual or linguistic frameworks and test their suitability for various practical purposes. These frameworks are tools, so we do not have to prove that they are correct. Nor do we have to agree on which ones to use. We just have to be clear enough to see what follows from what. Then a new result, whether it is a newly clarified concept or a new theorem is a new and permanent and positive addition to our stock of tools. And Carnap can offer the preceding three decades and more in logic as an example of the sort of continuing progress that he is describing. Logicians often disagreed about which systems to use, but they almost never disagreed about what were the results of one another's systems.

It will certainly be urged that Carnap's conception of philosophy here is very much less ambitious than the model of finding truth in one blazing insight. Certainly Carnap's conception *is* less ambitious, but it is also more realistic. The history of speculative philosophy, especially when unconnected to science, has been bleak. What good are lofty ambitions that in retrospect seem so often to have come to grief? Is it not better to put one foot in front of the other and make some real progress, however much that may seem, well, pedestrian?

I have said that on the new view, philosophy becomes conceptual engineering, and that the frameworks, concepts, and logico-mathematical structures are tools. But tools for what? What Carnap is thinking about most of the time in this connection is science. Carnap's philosophical programme is intended to be progressive in science in several different ways.

First, Carnap hopes that philosophers will clarify and refine some more or less traditional philosophical notions and that doing so will contribute to the progress of science. He would eventually call this task explication. Among the notions to be explicated are observation or protocol sentence, confirmation, probability, and theory. Actually this is more than a hope since Carnap is himself actively working on these issues. The principle of tolerance itself made its debut in the protocol sentence debate, and Carnap worked hard to disentangle various concepts of probability.

Second, sometimes logicians and mathematicians can devise and explore wholly new concepts that will prove useful in science. This was certainly true when the mathematicians developed non-Euclidean geometries in the nineteenth century. Another example is non-standard analysis as developed by Carnap's successor at UCLA, Abraham Robinson. This may well prove valuable, for example, in resolving problems with the measure function in certain probability theories. For Carnap, the ultimate test for the as-yet-unrecognized possibilities in wholly new concepts is the contribution they can make to the progress of science. The point, of course, is that there should be no a priori limits on what can be useful or what the source of help should be.

Third, philosophers can work directly with scientists. Philosophers of course are not the only ones who engage in conceptual clarification. Scientists do too. Clarifying concepts already in use in science is also explication. Disentangling lines of inference connecting observation and theory is, in effect, explication as well. So it is often productive to have scientific workers with different kinds of training, that is, to have philosophers and scientists standardly so called, working together on common problems. In recent decades as general philosophy of science has developed into philosophy of physics, philosophy of biology, philosophy of psychology, and so on, and on, and on, this kind of constructive fruitful interchange has become a reality.

Fourth and finally, philosophy can also promote and defend science itself, for example, by promoting and defending empiricism and the verifiability principle. Perhaps it is paradoxical, but as we saw, treating empiricism as a convention, as the principle of tolerance does, makes it stronger. And as we saw, treating the verifiability principle in the first instance as a proposal makes it easier to defend.

I have said that once philosophers adopt tolerance as a philosophical programme, they thereby abandon absolute correctness as the standard of their work in favour of pragmatic utility. I have also said repeatedly that

Carnap was most often thinking of progress in science as the measure of that utility. I did not say, nor would Carnap, that it was the only form of utility. Indeed, there is the vastly larger domain of social utility and social progress. I think Carnap was acutely aware of the enormous importance of his philosophical programme and of tolerance within it, for social progress. But while he was aware of the connection, he did not know what to do with it. So he said little and concentrated on what he knew best, namely the underlying work in logic and science. I do not know what to say either, except to stress the connection, so my remarks on this point will be brief.

Before leaving the topic of scientific progress, however, it is good to remember that scientific progress itself makes some indirect contribution to social progress. In a minor way this is effected by providing products or control that make life easier. Much more important, though, is that science helps to free us from superstition and free us from those habits of mind that keep us from thinking for ourselves.

But this tie through science between the programme of tolerance and social progress is indirect. The connection that I wish to point out now is immediate and powerful. There is no need to remind anyone that shortly before *Logical Syntax* was published, a government was elected in Germany that claimed to be the champion of certain traditional values, concepts, and institutions, all of which Carnap found repugnant. Central to tolerance is the idea that traditional social concepts such as duties, rights, property, the state, and marriage, can be refined and even replaced. Moreover, traditional social structures based on these concepts cannot be seen as uniquely correct but are themselves tools to be refined or even replaced. I think Carnap quite clearly saw his philosophical programme in opposition to traditional ways of thinking and as providing the basis for social reform. Furthermore, he quite clearly saw the explosive power of tolerance.

This is not to say that Carnap was social theorist or that he had a well worked-out social theory to offer. But he did have a conception of the enterprise according to which philosophers in following his programme could make a positive and constructive contribution to social progress.

Carnap's philosophical programme, then, is about as far from what I called the received unwisdom as can be imagined. It turns out to be a programme for positive and constructive work in philosophy, in science, and in society at large. Thus, it is a programme that to a very large extent has both its roots and its fruits in science. It is a programme generated by an idea that he simply did not have when he started writing the book. In this respect, as I said at the outset, the usual order is reversed. The initial stages of writing *Logical Syntax* produced the idea of tolerance, and that idea blossomed into the progressive programme. Of course, tolerance alone cannot guarantee results. But it can – gently – give us the strength to begin.

## Notes

1. See Awodey and Carus (2007, pp. 37ff.) and also Gödel (2003, pp. 346–57).
2. See Creath (1992, esp. pp. 150–3). See also Frege (1980, pp. 31–52) and Coffa (1991, pp. 135–7).
3. W.V.O. Quine, Letter to Carnap dated February 4, 1938, in Creath (1990, p. 241).
4. Carnap (1936–7, pp. 419–71, and pp. 1–40, esp.).

# Part IV

# Carnap, Empiricism, and the Principle of Tolerance

# 9
# From Tolerance to Reciprocal Containment

*Thomas Ricketts*

Carnap and Quine are usually presented as great philosophical opponents. And so they were. Their opposition should not blind us to the equally significant affinities in their philosophical outlooks. Burton Dreben encapsulates both the affinity and opposition at the end of his paper 'Quine':

> *Word and Object* is dedicated to 'Rudolf Carnap, teacher and friend'. To me the book gives the mirror-image of Carnap's philosophy: it shows how Carnap is transformed once his most basic assumption is dropped, namely the fundamental distinction between philosophy and science, between the analytic and the synthetic. (Dreben 1990, p. 88)

Encapsulations need to be unpacked, especially this one. What is the distinction between analytic and synthetic in Carnap's hands, and how does one give it up? Here, of course, we look first to Quine's critique of analyticity, but immediately run into difficulties. The problem is that Quine's specific criticisms do not meet up with Carnap's views, as Quine himself in effect notes in a prefatory paragraph to his paper 'Carnap and Logical Truth', Quine's contribution to the Carnap Schilpp volume:

> My dissent from Carnap's philosophy of logical truth is hard to state and argue in Carnap's terms. This circumstance perhaps counts in favor of Carnap's position.[1]

So let's begin with the first point. How are Carnap's and Quine's philosophies 'mirror-images'? Both philosophers lament obscurity, prize clarity, use logical notations and techniques to achieve clarity, and believe that philosophy in significant measure consists of what both philosophers call explication. Carnap, in his Schilpp volume autobiography, says:

> When I compared [controversies in traditional metaphysics] with investigations and discussions in empirical science or in the logical analysis of language, I was often struck by the vagueness of the concepts used

> and by the inconclusive nature of the arguments. I was depressed by disputations in which the opponents talked at cross purposes [...]. (Carnap 1963a, pp. 44–5)

For Carnap, the desired clarity is to be achieved through explication, which transforms 'a given more or less inexact concept into an exact one or, rather, in replacing the first by the second' (Carnap 1962, p. 3). Quine sounds much the same note in the last chapter of *Word and Object*, in the section entitled 'The Ordered Pair as Philosophical Paradigm':

> [In offering an 'explication' of an inadequately formulated 'idea' or expression, we] do not claim synonymy. We do not claim to make clear and explicit what the users of the unclear expression had unconsciously in mind all along. We do not expose hidden meanings, as the words 'analysis' and 'explication' would suggest; we supply lacks. We fix on the particular functions of the unclear expression that make it worth troubling about, and then devise a substitute, clear and couched in terms of our liking, that fills those functions.[2]

I think that the affinities and conflicts between Carnap and Quine rest in differences in their views of the relation of logical notations to colloquial language. These multifaceted differences are my topic here. My purposes are expository rather than evaluative. I'll be concerned to present a broad-gauged account, putting to the side various exegetical and philosophical issues that arise with respect to the interpretations I will sketch of Carnap and Quine. I want to begin with Carnap's principle of tolerance, which frames his understanding of explication and ties it to the analytic-synthetic distinction. I then turn to Quine's view of logic and how it seems to preclude Carnap's attitude of tolerance without really engaging Carnap's view. Finally, I set Quine's views on logic in the broader setting of his views on language and cognition in order to portray the genuine alternative he offers to Carnap's philosophy.

## I.

From the publication of *The Logical Syntax of Language* onwards, tolerance in logic is the central, determining feature of Carnap's outlook. In a 1935 address to the Paris Congress for Scientific Philosophy, Carnap describes three stages in the emergence of scientific philosophy. The first, Kant's contribution, is the rejection of speculative metaphysics in favour of theory of knowledge. The second stage is the rejection of the synthetic a priori in favour of a thoroughgoing empiricism. However, at this second stage, Carnap thinks that, '[...] *theory of knowledge* is [...] *an unclear mixture of psychological and logical elements*' (Carnap 1936a, p. 36). The third and final stage is the purification of epistemology that transforms it into the logic of science. Tolerance in logic is central to Carnap's conception of this transformation. In particular, the view of logic and its application encapsulated in the principle of tolerance will enable Carnap to demarcate psychological from logical elements within the purified epistemology that the logic of science delivers.

But how is Carnap's attitude of tolerance in logic possible? In the Foreword to *Logical Syntax*, Carnap notes with satisfaction the emergence of logic in *Principia Mathematica* as a rigorous science. He believes, however, that the further progress of logic has been impeded by the conviction that there is one, true, capital-L logic. Carnap's logic teacher, Gottlob Frege, held to this conviction.

Frege develops his logical notation, his *Begriffsschrift*, in order to use it to provide a codification of principles of demonstrative inference, a codification that makes possible the notationally secured rigour of gap-free proofs. Features of the contents of statements relevant to inference are notationally marked so that various inference modes can be characterized in notational terms. Frege's enterprise thus requires the replacement of redundancies and ambiguities of colloquial language with the uniformity of a *Begriffsschrift*. Any number of distinctions among colloquial sentences will be effaced by this replacement. Frege asserts that there is something that is preserved, what Frege calls the thought (*Gedanke*) that a sentence expresses. This notion of a thought in turn is elucidated by reference to the conception of judgement and truth that frames Frege's enterprise.

For Carnap to move beyond Frege, he will need an alternative to Frege's conception of language as expressing thoughts, the objects of judgement, standing in inferential relationships independently of us and our grasp of them.[3] Carnap adapts an alternative view of language from Hilbert. This Hilbertian view of language is crucial for understanding how tolerance is possible. Hilbert, in his metamathematical investigations, ignores the intended meanings, the intended use, of the signs of the formalisms he investigates, treating the formalisms as calculi which are constituted by notational rules. This treatment of formal languages as calculi is reinforced by Gödel's observation that the logical syntax of a formalism can be interpreted within arithmetic. Sentences can thus be identified with numbers; logical syntax becomes an application of arithmetic.

This view of languages as calculi is problematic in the context of Carnap's interest in the logic of science. When it comes to representing the testability of empirical theories, it won't do to think of languages as mere calculi. What is the relationship between a calculus and the actual or envisioned linguistic and non-linguistic activities of scientists? What is it to adopt a calculus as the language for science?

In §2 of *Logical Syntax*, Carnap says that languages are instances of calculi. This remark just pushes the question back. What is a language, and what makes a language an instance of a calculus? Carnap gives a broadly and non-dogmatically behaviourist answer to the first question:

> A language, as, e.g., English, is a system of activities or, rather, of habits, i.e., dispositions to certain activities, serving mainly for the purposes of communication and of co-ordination of activities among the members of a group.[4]

In §25 of *Logical Syntax*, Carnap explains how we coordinate a calculus with an actual or imagined language by correlating the well-formed formulas of the calculus with sentences of the language, i.e. specifying phonological or inscriptional realizations for the well-formed expressions of the calculus. He compares this coordination of calculus and language with the relationship between mathematical and physical geometry established by Reichenbachian coordinating definitions.[5] With this coordination in place, we can impose the logical syntax of the calculus onto the coordinated language like a grid. For example, modulo our coordination, we might note that the sentence uttered by such and such person at one time contradicts the one uttered by another person at a different time.

The possibility of thus coordinating languages and calculi is fundamental for Carnap's ambitions for the logic of science. The purification of theory of knowledge Carnap envisions proceeds via the explication of epistemological notions in logical terms. This enterprise thus requires the application of logical distinctions to languages in potential use. It is this application that links the austere, abstract classifications of logical syntax to the activities of scientists. To understand Carnap's crucial explication of the notion of empirical testability, there is a further aspect of the coordination of languages and calculi to consider – observation predicates.

In 'Testability and Meaning', Carnap maintains that the notion of an observation predicate is drawn from a biological-psychological theory of language (Carnap 1936–7, p. 454). Roughly speaking, a predicate is an observation predicate, if speakers of the language largely agree in their dispositions to affirm and deny the predicate to demonstrated items on the basis of their current observation of those items. Although Carnap does not explicitly discuss the matter, we can easily bring observation predicates into the coordination of calculi and potentially used languages. We segregate a group of predicates of the calculus – call them O-predicates – and pair them with the observation predicates of the coordinated language. We can now apply the machinery of logical syntax to the description of the revision of empirical theories on the basis of observation. Suppose an investigator has come to dissent from some sentence she previously held true. Having coordinated her language with a calculus, we logicians of science can represent this change as the rejection of a hypothesis on the basis of contradictions between formulas logically implied in the coordinated calculus by a theory the investigator held containing the rejected sentence, a factual (synthetic) sentence of the calculus, and observation sentences that appear in the investigator's protocol. In this example, we see how Carnap hopes to explicate epistemic notions in logical terms.

We also see in this example how Carnap makes a sharp distinction between psychology and epistemology as the logic of science. Via the coordination of an actual or imagined used language with a calculus, the acceptance and rejection of sentences can be represented as the epistemic evaluation of hypotheses on the basis of observation. Without the coordination of language and calculus, we have simply changes in linguistic dispositions, and so nothing that is epistemically or logically evaluable, i.e. no suitably precise vocabulary for epistemic or logical evaluation. In a revealing passage in the 1939 pamphlet, *Foundations of Logic and Mathematics*, Carnap describes the relation of languages to semantical systems – the successors to the calculi of *Logical Syntax*, in the following terms:

> The facts [about linguistic behaviour] do not determine whether the use of a certain expression is right or wrong but only how often it occurs and how often it leads to the effect intended, and the like. A question of right or wrong must always refer to a system of rules. Strictly speaking, the rules which we shall lay down are not rules of the factually given language B; they rather constitute a language system corresponding to B [...][6]

This view of the coordination of calculi with actual or potentially used languages provides the context for Carnap's principle of tolerance. Logicians of science elaborate and metamathematically investigate various calculi. Investigators may then select one or another calculus as representing the logical standards they intend to hold themselves to. Indeed, for Carnap, talk of truth and confirmation becomes suitably precise only in application to a calculus with its consequence-relation. We have then a sharp and principled distinction between epistemic evaluation of the acceptance and rejection of various sentences of a language in coordination with a calculus and the adoption of a calculus whose logical syntax makes such evaluation possible. No language-relative notion of truth or correctness applies to the choice of a calculus as the language of science. Thus, Carnap advocates an attitude of tolerance in logic. The choice of a calculus as the language for science thus resembles the choice of a set of rules for a game. The rules define what is permitted in the course of the game. But no such question of legitimacy applies to the choice of these rules themselves. The attitude of tolerance combined with Carnap's conception of the relation of the description of a calculus to linguistic behaviour thus opens up the prospect of a rigorous, scientific philosophy, while enforcing a sharp distinction between philosophy as the logic of science and substantive science itself.

This attitude of tolerance in turn frames Carnap's reconciliation of logicism with empiricism. Carnap attempts to characterize in logical terms relationships between observation predicates and other non-logical predicates which will ensure that theories stated using the non-logical signs will have a rich array of testable consequences. Such a language is an empiricist language. To be an empiricist is then to resolve to restrict one's choice for the language of science to empiricist languages.[7] The L-consequence relation of a calculus will standardly determine some formulas to be L-valid and others to be L-contravalid, L-true and L-false as Carnap will later say. In an empiricist language, the L-validity and L-contravalidity of these sentences is an artefact of the L-consequence relation that secures the empirical testability of theories in the language. Carnap speaks of these L-determinate sentences of an empiricist language as notational auxiliaries to the L-indeterminate sentences, the synthetic sentences, of the language (Carnap 1935b, p. 34). In *Logical Syntax*, Carnap shows us how to construct calculi each of whose purely mathematical sentences is L-determinate (i.e. either L-valid or L-contravalid). In this way, mathematics becomes part of the logic built into a language. In an empiricist language, the mathematical sentences thus play the role of notational auxiliaries. We see here how the attitude of tolerance builds in a particular understanding of the analytic-synthetic distinction.

Carnap's shift from syntax to semantics does alter the form that the definitions of L-consequence and L-truth take for empiricist languages and prompts changes in his expository rhetoric. Nevertheless, I maintain that the shift to semantics does not change the basic Hilbertian view of language. To understand this, we need to consider what Carnap's shift to semantics in the first instance involves. In *Logical Syntax*, to build classical mathematics into a calculus, Carnap must employ very strong resources in his syntax language (metalanguage) definitions of L-valid and L-contravalid. Roughly speaking, he defines in his syntax language predicates 'valid' and 'contravalid' that function as bivalent truth- and falsity-predicates over the purely logical-mathematical formulas of a calculus (as opposed to those to be used for the expression of empirical science). Because these definitions use only logical and mathematical vocabulary of the syntax language, Carnap counts the definitions of 'valid' and 'contravalid' syntactic, even though he is well aware from Gödel's incompleteness theorem that they cannot be captured by derivability in any formal system. He thus takes himself to have shown how to reproduce in syntactic terms the true–false distinction over the purely logical-mathematical formulas of a calculus that builds in classical mathematics.[8] Of course, we can extend these definitions of 'valid' and 'contravalid' by formalizing some portion of empirical science. The description of the formalization in the syntax language will, of course, be in syntactic terms. Nevertheless, Carnap reasonably thinks that no formalization of empirical knowledge of any significant scope in a calculus will settle the truth or falsity of all the empirical sentences, or will extend the true–false dichotomy from the logical to the non-logical sentences of the calculus. For Carnap thinks there is no way of using a purely logical-mathematical metalanguage to extend the definitions of 'valid' and 'contravalid' to determine, for example, the truth-value of all the singular empirical sentences assigning a colour to a space-time coordinate. This is the point Carnap makes in *Logical Syntax*, when he says:

> *For truth and falsehood are not proper syntactical properties*; whether a sentence is true or false cannot generally be seen by its design, that is to say, by the kinds and serial order of its symbols. [This fact has usually been overlooked by logicians, because, for the most part, they have been dealing not with descriptive but only with logical languages, and in relation to these, certainly 'true' and 'false' coincide with 'analytic' and 'contradictory', respectively, and are thus syntactical terms.] (Carnap 1934d/1937, §60b, p. 216)[9]

Tarski shows how a bivalent truth-predicate can be defined over both the logical-mathematical and the descriptive sentences in languages of interest to Carnap in a metalanguage with the logical-mathematical resources of Carnap's syntax languages. Tarski's success comes at a price: a Tarskian truth-definition must use descriptive predicates *in the metalanguage* to specify satisfaction conditions for object-language descriptive predicates. The truth-predicate Tarski defines does not then count as syntactic by Carnap's standard. Still, the use of descriptive predicates to state satisfaction conditions for descriptive predicates is innocent enough. In particular, Tarskian truth-definitions exploit no information about the extensions of descriptive predicates they use. This use of descriptive predicates in the metalinguistic specification of a formal language is, in the first instance, what Carnap's shift from syntax to semantics amounts to. Carnap's leading idea now is that the L-truths of a formal language are those sentences whose truth is a logical consequence *in the metalanguage* of a truth-definition for the formal language. For every sentence S in the formal language, the truth-definition will logically imply a biconditional:

S is true in L if and only if p.

The L-truths are those sentences for which the truth-definition also logically implies:

S is true in L.[10]

Of course, this characterization of L-truth for a formal language does not give us a *definition* of L-truth, as it mentions logical consequence in the metalanguage. Carnap takes this characterization to provide an adequacy condition for a definition of L-truth stated *in* the metalanguage (see Carnap 1942, pp. 83–4). In *Introduction to Semantics*, Carnap considers formalisms that include a type-stratified quantificational logic. He proposes that the logical vocabulary be stipulated by enumeration, assuming that truth-functional connectives, quantifiers, and variables will be included in it. He further assumes that there will be a type of variable corresponding to every primitive descriptive expression. Following Tarski, Carnap suggests that a formula in the formalism is L-true just in case its universal closure with respect to any primitive descriptive vocabulary is true. Thus, for a range of semantical systems, L-truth can be defined in terms of truth.[11] Later, Carnap suggests appending to a truth-definition for an extensional formal language a model-theoretic definition of L-truth, tailored to deliver the desired results.[12] Through all this, I claim that the idea of giving formalisms application in the logic of science by coordinating the formulas of a formalism with the sentences of a potentially used language remains in place.

## II.

In the summer of 1954, Quine and Carnap exchanged letters about Quine's paper, 'Carnap and Logical Truth'. Carnap queried Quine in preparing his reply to this paper:

> Now there is a point where I should like some clarification of what you mean so that my reply can be more specific. [...] The question is which of your discussions are meant to refer to (a) natural languages, and which to (b) codified languages, language systems based on explicitly formulated rules. [...] The distinction is of great importance for my discussion, because from my point of view the problems of analyticity in the two cases are quite different in their character. (Carnap to Quine, 15 July 1954, Creath 1990, p. 435)

Quine replied:

> It is indifferent to my purpose whether the notation be traditional or artificial, so long as the artificiality is not made to exceed the scope of 'language' ordinarily so-called, and beg the analyticity question itself. [...] The languages I am talking about comprise natural languages and any (used, or interpreted) artificial notations you like, e.g. that of my *Mathematical Logic* plus extra-logical predicates. They are not uninterpreted notations. Each predicate has its unique extension, and correspondingly for the logical signs [...]. (Quine to Carnap, 9 Aug. 1954, Creath 1990, pp. 437–8)

We should sympathize with Carnap's puzzlement. Quine does not fault the technical features of definitions of L-validity that Carnap has provided for various extensional formalisms. Rather, Quine questions the point of Carnap's constructions:

> Obviously any number of classes K, M, N, etc. of statements of $L_0$ can be specified for various purposes or for no purpose; what does it mean to say that K, as against M, N, etc., is the class of the 'analytic' statements of $L_0$? (Quine 1951/1961, p. 33)

Some of Quine's challenges here fail to appreciate the deflated philosophical ambitions of Carnap's logic of science, the way in which logically explicated epistemic vocabulary purifies and replaces older, vaguer notions. For example, there is no appeal in Carnap to truth by convention, and no weight, I argue, is put on any notion of truth solely in virtue of meaning. For Carnap, vague appeals to meaning, in a sense in which meaning is linked to understanding, should be replaced by reference to the rules that define a formal language. From Carnap's perspective, Quine's critique must then come down to the basis for coordinating the transformation rules of a calculus with a language. The problem is most pressing in the case of Carnap's meaning postulates on account of the formally irregular ways they mix, as Carnap would put it, logical and descriptive vocabulary. But, as I have presented Carnap's views, his conception of the application of calculi in the logic of science is not premised on any basis in hypothetical or actual linguistic behaviour for coordinating the transformation rules of a calculus with a group's used language. Coordination of the formation rules is enough to give us logicians of science the logical vocabulary to voice logically explicated epistemic evaluations of a group's acceptance and rejection of hypotheses on the basis of observation. Carnap, as always, wants to convert fruitless wrangling into a productive discussion, and so strives to answer Quine. Nevertheless, I take very seriously Carnap's disclaimer in his Schilpp volume reply to Quine:

> As I now understand Quine, I would agree with his basic idea, namely, that a pragmatical concept, based upon an empirical criterion, might serve as an explicandum for a purely semantical reconstruction and that this procedure may sometimes... be a useful way of specifying the explicandum. On the other hand, I would not think that it is necessary in general to provide a pragmatical concept in order to justify the introduction of a concept of pure semantics. (Carnap 1963b, p. 919. See also Carnap 1955/1956, p. 234)

How do things look to Quine? How can he so blithely ignore the distinction between colloquial languages and formal languages?

Quine takes logical notation to be a part of, an extension of, colloquial language. In general, the systematic pursuit of knowledge involves the introduction of new or adjusted terminology and other notational innovations in the interest of clarity and simplicity. The new and adjusted terminology is, of course, initially explained in antecedently familiar terms. The development and use of logical notation – truth-functional connectives in combination with quantifier-variable notation for generality – is just more such linguistic innovation. It is continuous with the use of letters to replace pronouns in colloquial mathematics and with the use there of parentheses to indicate the order of arithmetical operations. Logical notation thus becomes a part of our colloquial language to be used on occasion in the interests of clarity in place of other parts of our language. As I've already mentioned, Quine does not presuppose any synonymy, any identity of content or sense, between the sentences of colloquial English and the sentences containing the devices of quantificational notation that on occasion replace everyday language.

Quine's logical notation – truth-functional connectives and quantifiers with their variables in singular term positions – comprises a partial notation for discourse on any topic. In taking this view of quantificational notation, Quine rejects Hilbert-inspired reference to uninterpreted formalisms in logic in favour of semantic ascent.[13] Quine construes the 'p's' and 'q's' that appear in sentence positions in truth-functionally compound formulas and the 'F's' and 'G's' that appear in predicate positions in quantified formulas to be schematic letters – placeholders, blanks marking positions for sentence and predicates, respectively. The formulas in which they occur are not 'uninterpreted sentences', but diagrams of the forms of sentences, skeletal sentences that are fleshed out into sentences by the uniform replacement of schematic letters by sentences or predicates. A chief advantage of quantificational notation are the few, uniform constructions it provides for building up sentences ultimately from primitive predicates so that the truth or falsity of the sentence is transparently fixed by the extensions of the predicates. We have then for sentences in quantificational notation a tractable notion of form that yields a clear notion of logical validity: a sentence is logically valid, if the sentences sharing its form are all true. Logical laws generalize over the forms of sentences with respect to truth, as in

Every disjunction of a sentence with its negation is true.

Moreover, it turns out that there is a complete proof procedure for establishing the validity of quantificational forms.

Quine's approach to logic then relies heavily on the notion of truth and an associated notion of reference or satisfaction for predicates. Quine finds the notion of *truth* embodied in the colloquial predicate 'true' to be largely unproblematic.[14] He notes the systematic material equivalence between sentences and ascriptions of truth to those sentences exhibited by the Tarski paradigm:

'________' is true if and only if ________.

This systematic material equivalence, Quine urges, makes the predication of truth to a sentence as clear as the original sentence, and generally serviceably as a paraphrase of it. The predicate 'true' is thus used without any implicit reference to semantic rules:

> Where it makes sense to apply 'true' is to a sentence couched in the terms of a given theory and seen from within the theory, complete with its posited reality. *Here there is no occasion to invoke even so much as the imaginary codification of scientific method.*[15]

This clarity in turn suits the predicate 'true' for use in semantic ascent to state a precise, illuminating (in part, because extensional) definition of logical implication for sentences couched in quantificational notation. Once quantificational notation is entrenched in language, familiar laws of quantificational logic are obvious: their truth is, practically speaking, unquestionable. For Quine, Carnap's attitude of tolerance in logic makes no sense.

A Carnapian will not be moved from tolerance by Quine. The most objectionable part of Quine's view is the alleged use of a colloquial truth-predicate to characterize logical validity. It is Quine's use here of the predicate 'true', a use that Quine presents as continuous with the application of this predicate to colloquial sentences, that gives his presentation of logic its intolerant, dogmatic air. To appreciate what a Carnapian might find amiss here, let's go back to consider an issue I have neglected. In what sense does Carnap's view of formal languages remain 'Hilbertian' after the switch to semantics, after the admission of descriptive vocabulary into the descriptions of formal languages?

The answer in brief is that the definition of 'true-in-S' in the description of a semantical system S is, officially speaking, a stipulative definition, just like the definition of 'valid-in-S' was in *Logical Syntax*. We might as well call it the definition of a new predicate, 'T-in-S'. Of course, Carnap has read his Tarski. He recognizes that we may give a truth-definition for a language in a metalanguage that includes it. But this means: we may state a truth-definition for a semantical system S in a used language that we coordinate with a semantical system that includes S. In this case, we will recognize the sentences that occur on the right-hand side of the T-sentences the definition implies to be translations of the sentences designated on the left-hand side, because that's the way we set things up. But this case is not in any way privileged. The description of a semantical system is one thing; its coordination with a used language is something else. Of course, the truth-predicate stipulated in describing a semantical system coordinated with a used language may explicate that language's colloquial truth-predicate.

How then does my Carnapian view Quine's exposition of logic? I followed Quine's expository preference in *Philosophy of Logic* in characterizing logical validity for a quantificationally regimented segment of language by means of semantic ascent in terms of lexical substitution. For a suitably rich segment of language, e.g. one that includes arithmetic, this definition is equivalent to the standard model theoretic definition of quantificational validity.[16] This characterization does not use the colloquial predicate 'true'; it uses a set-theoretically defined notion of *truth in a model*. My Carnapian might seize on this characterization as the properly precise one, and would, in any event, reject Quine's use of a colloquial truth-predicate in serious logic. He will accordingly view Quine's semantic ascent characterization of validity as masking what is better viewed as a pragmatically motivated adoption of first-order quantificational logic as the logic of science. So viewed, Quine offers neither a challenge nor a clearly conceived alternative to Carnap's position.

## III.

With Carnap we have the coordination of semantical systems with actual or potentially used languages. A truth-predicate in application to a semantical system is, officially speaking, introduced by stipulative definition into the description of the system. With Quine we have the addition of logical notation into colloquial language and the regimentation of colloquial sentences into this notation. The colloquial truth-predicate is applicable to sentences in logical notation. What is at issue here is Carnap's Hilbertian view of language and the distinction it underwrites between adopting a logic and applying the adopted logic in the epistemic evaluation of hypotheses – the distinction between choosing the rules and playing the game, which is central to Carnap's philosophical project. However, as far as we have gone, Quine's differences with Carnap appear too slight to support a genuine alternative to Carnap's approach. To understand the substantive differences between Carnap and Quine, I want to set the differences noted in the last section in the context of their respective views of empiricism.

Carnap holds that we gain a suitably precise vocabulary for evaluating scientific theories only by adopting a formal language as the logic of science. The adoption of a formalism is not similarly subject to evaluation. Carnap also holds that substantive scientific theories should be evaluated by reference to their observational consequences. But why rely on observation as the evaluative touchstone for knowledge? As I have already observed, for Carnap empiricism is a proposal, an attitude, not a thesis:

> It seems to me that it is preferable to formulate the principle of empiricism not in the form of an assertion – 'all knowledge is empirical' or 'all synthetic sentences that we can know are based on (or connected with) experiences' or the like – but rather in the form of a proposal or requirement. As empiricists, we require the language of science to be restricted in a certain way; we require that descriptive predicates and hence synthetic sentences are not to be admitted unless they have some connection with possible observations, a connection which has to be characterized in a suitable way. (Carnap 1936–7, p. 33)

Empiricism itself is an ultimate preference, a cognitive value that cannot itself be termed correct or incorrect. Quine thinks otherwise. In the opening section of 'The Scope and Language of Science', a paper written in 1954 at the height of the analyticity debate with Carnap, Quine asserts that science tells us that 'in our knowledge of the external world we have nothing to go on but surface irritations', a point Quine frequently repeats in subsequent writings.[17] What does this alleged truism come to? We voice the knowledge thus gained by use of the sentences of our language. The dispositions to assent to some sentences and dissent from others that underlie this use arise and change over time mainly as a result of our sensory stimulations. Here we have a genuine truism, but without the epistemically loaded rhetoric of *what we have to go on*. What does this rhetoric come to in Quine's hands?[18]

Quine says that the origin of our knowledge in sensory stimulation raises the question:

> Whence our persistence in representing discourse as somehow *about* a reality, and a reality beyond the irritation? . . . Whence the idea that language is occasionally descriptive in a way that other quiverings of irritable protoplasm are not? (Quine 1957/1976, p. 230)

Quine addresses this question in his treatment of language acquisition – treatment that begins in earnest with his 1958 paper 'Speaking of Objects', continues in chapter 3 of *Word and Object*, and gets its most extended development in the 1973 book *Roots of Reference*. This account will lead us to an understanding of Quine's epistemic rhetoric.

Infants take the first steps in acquiring a language when they learn to respond with words to distinctive kinds of stimulations caused by their surroundings: to respond with 'Mama' to stimulations caused by a particular person, with 'Water' to simulations caused by that liquid; with 'Dog' to stimulations arising from dogs. Quine insists that this early step in language acquisition does not deliver words that refer to things. To paraphrase Quine, the infant has no grasp on the distinctions which we adults might voice by 'Mama again', 'More water', 'Another dog'. For the child, it is, so to speak, 'More Mama', 'More water', 'More dog' (see Quine 1960, p. 92).

The child acquires our conception of enduring things, as the child masters those words and grammatical devices that embody that conception, what Quine calls our referential apparatus. So the child must learn to use general terms, especially the individuative ones like 'dog' that divide their reference. The child must learn to use them with demonstratives to form singular terms ('That dog is nice.'). The child should learn as well constructions such as plural predication or singular predication with indefinite singular terms as subjects to express generalizations ('Dogs are animals.'; 'Every dog buries bones.'). The child must also master expressions of identity, difference, and number in connection with individuative general terms ('That dog is not Fido.'; 'Three dogs are playing in the backyard.'). Quine thinks that acquisition of the conception of enduring things is fully-fledged only with the mastery of the relative clause construction. For this construction expands the stock of general terms far beyond those like 'dog' whose acquisition began, or might have begun, as a prompted verbal response to current stimulation,[19] i.e. beyond those capable of functioning holophrastically as observation reports.

Quine understands the child's mastery of referential apparatus in broad and vague behavioural terms. For a child to have mastered the referential apparatus is for the child to have the ability to use and respond appropriately to a broad range of sentences involving the referential apparatus so that the child's uses and responses pass muster among adult speakers. Quine gives almost no details about the emergence of this mastery. He says things like:

> The child learns this apparatus by somehow getting a tentative and faulty command of a couple of its component devices, through imitation or analogy perhaps, and then correcting one against the other, and both against the continuing barrage of adult precept and example, and going on in this way until he has a working system meeting social standards.[20]

How then does Quine think of himself as answering the question he raises in 'Scope and Language' about the referential character of language? I want to approach matters via a second question: Why does Quine focus on the words and constructions he calls the referential apparatus? What motivates grouping these words and constructions together?

One might expect from Quine some characterization of reference, of the linguistic function of referring terms, perhaps in contrast to predication. The referential apparatus of the language are then the ways that language has of fulfilling this function. That is what Strawson expects, and in his classic paper, 'Singular Terms and Predication', chides Quine for not providing it. Quine demurs. He thinks there is no account of what reference is, of the function of referring terms that is independent enough of the referential apparatus of our language to serve as a basis for identifying that apparatus.

Quine begins chapter 3 of *Word and Object*, 'The Ontogenesis of Reference', with the remark:

> When in English we decide whether a term is meant to refer to a single inclusive object or to each of various of its parts, our decision is bound up with a provincial apparatus of articles, copulas, and plurals that is untranslatable into foreign languages save in traditional or arbitrary ways undetermined by speech dispositions. Toward understanding the workings of this apparatus, the most we can do is examine its component devices in relation to one another and in the perspective of the development of the individual or the race. (Quine 1960, p. 80)

He accordingly replies to Strawson:

> In a sense, thus, Strawson is right in saying that I explain not the distinction between general and singular, but only the form of signaling it. He would be wrong in supposing that I thought I had or should have done more.[21]

This response only makes the question of the identification of the referential apparatus all the more pressing. If Quine does not have something to say to motivate his identification of the referential apparatus, then it becomes completely opaque how Quine's discussion of language acquisition can have the relevance he claims for it.

Quine's referential apparatus consists of those grammatical devices that correspond to features of logical notation. For Quine, the interest and importance of logic is bound up with its application in science. In science, investigators test hypotheses by deriving observationally testable consequences from those hypotheses taken together with background knowledge and assumptions. One wants then some account of the relation of implication that empirical theory testing makes salient. This account is desirable, not so much to settle disputes about what implies what, as for its own sake. We should want an account of implication that will back up the casual appeals to it of working investigators. One wants the account as itself a part of science.

Quine urges that quantificational logic, despite its austerity, is sufficient to provide the implicational bridges linking bodies of theory with empirical check points. It then has good claim to the title 'logic of science'. Quine's account of quantificational implication talks of truth and reference (satisfaction) with respect to the formulas of logical notation, as noted in the previous section. It is from the perspective of this account of implication, and the role that reference plays in it, that Quine identifies the referential apparatus of colloquial language. The referential apparatus consists of those grammatical devices that are both used to explain quantifier-variable notation and that are via regimentation replaced by it. This relationship motivates grouping these devices together under this rubric 'referential apparatus', and gives this apparatus, and with it, the notion of reference, its salience and interest.

We can then look at matters in two ways. On the one hand, quantificational notation represents a refinement, a regimentation, and an alternative to certain words and constructions present in colloquial language. On the other hand, it is through the role of this notation in the theory of implication that we go on to identify the referential apparatus of colloquial language. More than this, it is the role that the concepts of truth and reference play in the theory of implication and the role that implication plays in theory testing that together motivate Quine's identification of mastery of the referential apparatus with acquisition of a conceptual scheme of enduring things. The difficult thing to grasp here is that Quine makes neither of these two moments prior to the other. This is why he refuses to follow Carnap in sharply distinguishing colloquial language from formalisms. Let me explain.

Let's go back to Quine's attitude toward empiricism, his view that 'our information about the world comes only through impacts on our sensory receptors' (Quine 1990b/1992, p. 19). A person's theory, let's say, are the standing sentences of her language to which she is disposed to assent, if queried, unprompted by other current stimulations. After language learning is well along, modifications in a person's theory are largely produced by sensory stimulations. As Quine presents matters, these stimulations activate dispositions to assent to or dissent from observation sentences.[22] This in turn produces changes in dispositions to assent to and dissent from standing sentences linked to the observation sentences. Activation of a disposition to assent to 'That's a swan that's not white.' might convert a disposition to assent to 'All swans are white.' into a disposition to dissent. We don't yet have anything I want to dignify with the title *information*, but we do have in observation sentences, sentences to assent to and dissent from which are keyed to current stimulations, and so to changes in a person's sense organs that in turn typically reflect changes in the person's surroundings.

I have described Quine's view of the application of logic to make sense of the concept of implication invoked in theory-testing. Seeing quantificational notation as continuous with colloquial language enables us, via regimentation, to discern logical relationships over the sentences of colloquial language. Talk here of logical links, of implication, brings the concepts of truth and reference into play. In the swan case I just described, we can see dissent to a generalized conditional prompted by assent to the negation of an instance. The application of concepts of truth and reference to the sentences of our theory supports talk of the theory's being about things. In particular, application to observation sentences of concepts of truth and reference enables us to think of them as describing states of the things correlated in a rough and ready way with the stimulations which prompt assent to them. This is what supports Quine's talk of sensory stimulations carrying information about the world, his talk of those stimulations as evidence for theories. I have noted the role Quine sees the activation of assent-dissent dispositions to observation sentences by current stimulations to play in the revision of theory. Reflecting on matters in this way, the epistemic norm encapsulated in the hypothetico-deductive method emerges as a truism: in order to have true theories, we should then evaluate those theories by reference to what we learn about our surroundings from sensory stimulations, by reference to observation sentences.

Quine accordingly emphasizes the possibility of regimenting theories so that they imply generalized conditionals, observation categoricals, dissent from which would be prompted by assent to negations of their instances. As he views matters, this possibility shows that there is no sharp distinction between the genetic account of assent-dissent dispositions and the epistemic norm encapsulated in the hypothetico-deductive method. Not at the level of generality and abstractness Quine is working at. Application of the hypothetico-deductive method, broadly speaking, leads to the genetic account of assent-dissent dispositions Quine sketches. Science contains epistemology, as Quine puts it. That account reveals that something approximating the hypothetico-deductive method was operative all along in shaping our assent-dissent dispositions. Something like the hypothetico-deductive method is a natural necessity, not an option to be selected. Epistemology contains science.

We can now see more clearly the significance of Carnap's Hilbertian conception of formal languages. This conception enables him to make a sharp distinction between formal languages and colloquial language, and thus to balance tolerance in logic, empiricism in epistemology, and a principled distinction between psychological investigations of cognition and the development and articulation of epistemic norms. Quine's refusal of a sharp distinction between formal and colloquial language marks his rejection of the kind of principled distinction between the descriptive and the normative on which Carnap's philosophy is premised.[23]

## Notes

1. Quine (1963, p. 385). The prefatory paragraph is omitted from the reprinting of the paper in Quine (1976).
2. Quine (1960, pp. 258–9). In a footnote on p. 259 Quine aligns his view of explication with Carnap's, citing Carnap's discussion of explication in Carnap (1956, p. 8). See also Quine's discussion in Quine (1960, §33). For an earlier expression of this attitude, see Quine (1953c/1976, p. 150).
3. For further discussion of the contrast between Frege and Carnap, see Ricketts (2004). Even before adoption of the principle of tolerance, in *Aufbau*, Carnap had in effect rejected a Fregean conception of thoughts. This is evident in his

acceptance of constitution systems with different bases which, nevertheless, still give expression to the same body of knowledge. Note also his characterization of Fregean sense in markedly un-Fregean psychological terms in Carnap (1928a/2003, §§44, 50, and 51).

4. Carnap (1939a, p. 3). In Carnap (1934d/1937, §62), Carnap speaks of a translation from French into German as being in accord with sense (*sinngemäß*), and comments that this means only that 'the translation should stand in agreement [*Einklang*] with the historically familiar speech habits [*historische vorliegende Sprechgewohnheiten*] of French- and German-speaking people.' I don't see any compelling reason to think that Carnap fundamentally alters this view of used languages later in his career, but the matter requires a discussion of Carnap's attempt to explicate analyticity with respect to meaning postulates in pragmatic terms.
5. Carnap says nothing about what makes an utterance-type, a series of phonemes, a sentence of a group's language. I believe he would be happy to avail himself here of the account of grammar construction Quine sketches in Quine (1953b, §§2–3).
6. Carnap (1939a, pp. 6–7). See also Carnap (1942, p. 14), and Carnap to Quine, 21 Jan. 1943 in Creath (1990, p. 309). In Carnap (1935b), Carnap says of the distinction between pure mathematics and statements of empirical (factual) science, 'While in their psychological character, there is only a difference of degree and not a principled difference between the two fields, from a logical point of view a precise and principled difference can be demonstrated.'
7. For non-cognitive character of Carnap's commitment to empiricism, see Carnap (1936–7, p. 33). I return to this passage below.
8. So, for Carnap, the realm of the syntactic extends well beyond the realm of the recursively enumerable.
9. This is the reason why Carnap in *Logical Syntax* maintains that the notion of truth, as opposed to logical truth (analyticity), is not a syntactic notion and so is irrelevant to logic and its application to the logic of science. For further discussion of this point and Carnap's reception of Tarski's technique for defining truth, see Ricketts (1996), especially section ii.
10. After the shift to semantics, just as before, Carnap assumes in his metalanguage a strong relation of logical consequence that outstrips derivability in any formal system.
11. See Carnap (1942, pp. 86–7). Carnap here follows Tarski's characterization of logical consequence presented at the 1935 Paris Congress of Scientific Philosophy. See Tarski (1936b). This is one of several approaches Carnap considers in §16 of *Introduction to Semantics*, and the one he pursues in later writings. See Carnap (1954/1958). Carnap notes (1942, p. 87) that this characterization of L-truth will not capture L-truths determined by relations of meaning among descriptive vocabulary.
12. See Carnap (1963b, pp. 900–1). Carnap notes that the model theoretic characterization of L-truth 'can be stated in general semantics'.
13. See Quine (1960, p. 273). Note also the brusque dismissal of uninterpreted/disinterpreted formalisms in Quine (1963, §iv).
14. Of course, the semantical paradoxes require some adjustment of the predicate 'true' in its impredicative applications.
15. Quine (1960, p. 24, my italics). Quine had been discussing Peirce's identification of truth as the product in the limit of the continuous application of scientific

method to experience. I think his point is applicable as well to Carnap's restriction of suitably precise attributions of truth to the sentences of formal languages.

16. Quine discusses this point in Quine (1970) and (1986, pp. 53–5).
17. Quine (1957/1976, p. 230). See Quine (1960, p. 4); (Quine 1969b, p. 75); Quine (1990b/1992, p. 19).
18. In Quine (1990b/1992, p. 19), Quine says, 'our information about the world comes only through impacts on our sensory receptors'. Given Quine's rejection of intensional notions of content, what can he mean here by 'information'?
19. See Quine (1981b, pp. 4–8). See also Quine (1960, pp. 110–14).
20. Quine (1973, p. 84). Cf. Quine (1960, p. 93). The added details Quine delivers in Quine (1973) are concentrated on the initial, irreferential stage of language learning, and on the mastery of the relative clause construction, which, as Quine presents it, presupposes that the mastery of simple individuative terms like 'dog' and the devices associated with singular and plural predication are well-advanced.
21. Quine (1969c, p. 320). Quine goes on to note the uncomfortable parallel between Strawson's criticism of him and his of Carnap's view of analyticity.
22. Quine introduces his notion of an observation sentence in Quine (1960, pp. 40–6). For subsequent changes, see Quine (1990b/1992, pp. 2–6), and Quine (1993).
23. I am grateful for discussion on the topics of this chapter with André Carus, Juliet Floyd, Michael Friedman, Peter Hylton, and especially Warren Goldfarb. I also benefited from discussion of earlier presentations of these ideas at the University of Missouri at St. Louis and the University of Chicago.

# 10
# Tolerance, Intuition, and Empiricism

*Michael Friedman*

Although Carnap had defended a modified version of the Kantian conception of pure spatial intuition and geometry in his doctoral dissertation, *Der Raum* (1922), it seems that he had abandoned both pure intuition and the synthetic a priori while he was working on *Der logische Aufbau der Welt* in 1924–5, and this was certainly the case when he became a leading member of the Vienna Circle in the mid to late 1920s. In a well-known passage in his 'Intellectual Autobiography' (1963) Carnap describes how the characteristic Vienna Circle doctrine of the analytic character of all logical and mathematical truths, based on the Frege–Russell reduction of mathematics to logic and Wittgenstein's conception of tautology, allowed them to make a major advance over earlier forms of empiricism:

> What was important in this conception from our point of view was the fact that it became possible for the first time to combine the basic tenet of empiricism with a satisfactory explanation of the nature of logic and mathematics. Previously, philosophers had only seen two alternative positions: either a non-empiricist conception, according to which knowledge in mathematics is based on pure intuition or pure reason, or the view held, e.g., by John Stuart Mill, that the theorems of logic and of mathematics are just as much of an empirical nature as knowledge about observed events, a view which, although it preserved empiricism, was certainly unsatisfactory. (1963a, p. 47)

Indeed, this rejection of pure intuition and the synthetic a priori in favour of the view that all logico-mathematical truth is analytic and has no factual content quickly became definitive of what Carnap and the Vienna Circle meant by their empiricism.

In the late 1920s and early 1930s, however, the Circle became involved with the 'crisis' in the foundations of mathematics precipitated by Brouwer's development of a Kant-inspired version of 'intuitionism' concerning the objects of arithmetic and analysis and Hilbert's development of his proof-theory in response to Brouwer. In particular, Brouwer gave a famous lecture in Vienna in 1928, and, as Carnap further explains in his autobiography, the Circle was appropriately impressed:

> In the Circle we also made a thorough study of intuitionism. Brouwer came to Vienna and gave a lecture on his conception, and we had private talks with him. We tried hard to understand his published or spoken explanations, which was sometimes not easy. The empiricist view of the Circle was of course incompatible with Brouwer's view, influenced by Kant, that pure intuition is the basis of all mathematics. On this view there was, strangely enough, agreement between intuitionism and the otherwise strongly opposed camp of formalism, especially as represented by Hilbert and Bernays. But the constructivist and finitist tendencies of Brouwer's thinking appealed to us greatly. (1963a, p. 49)

One way to understand the problem with which the Circle was now faced, therefore, is how to acknowledge the evident strengths of Brouwer's viewpoint without becoming entangled with a 'non-empiricist' commitment to pure intuition. And what this ultimately means is that any account of mathematics we may offer – whether intuitionist or classical – must depend on the analytic or essentially contentless character of all logico-mathematical truth.

Carnap's solution to this problem, of course, is *The Logical Syntax of Language* (1934). In conformity with the basic metamathematical method of Hilbertian proof-theory, we view any formulation of logic and mathematics as a syntactically described formal system, where the notions of well-formed formula, axiom, derivation, theorem, and so on can all be syntactically expressed. In light of Gödel's recently published incompleteness theorems, however, we do not pursue the Hilbertian project of constructing a proof of the consistency of classical mathematics using finitary means acceptable to the intuitionist. Instead, we formulate both a formal system or calculus conforming to the strictures of intuitionism (Carnap's Language I, a version of primitive recursive arithmetic) and a much stronger system adequate for full classical mathematics (Carnap's Language II, a version of higher-order type theory over the natural numbers as individuals). For both systems, moreover, we define a notion of logical truth (analyticity) intended syntactically to express their essential independence from all factual content. Finally, and most importantly, Carnap promulgates the principle of tolerance: *both* types of system should be syntactically described and investigated, and the choice between them, if there is one, should then be made on practical or pragmatic grounds rather than prior purely philosophical commitments.

Directly following his discussion of intuitionism, Carnap presents a very clear and succinct description of the *Syntax* view:

> According to my principle of tolerance, I emphasized that, whereas it is important to make distinctions between constructivist and non-constructivist definitions and proofs, it seems advisable not to prohibit certain forms of procedure but to investigate all practically useful forms. It is true that certain procedures, e.g., those admitted by constructivism or intuitionism, are safer than others. Therefore it is advisable to apply these procedures as far as possible. However, there are other forms and methods which, though less safe because we do not have a proof of their consistency, appear to be practically indispensable for physics. In such a case there seems to be no good reason for prohibiting these procedures so long as no contradictions have been found. (1963a, p. 49)

And, as we know, the principle of tolerance then becomes absolutely central to Carnap's philosophy from this point on.

It is striking, then, that Carnap's views on syntax, intuition, and tolerance have been challenged by E. W. Beth, in an important paper in the Carnap Schilpp volume devoted to *Logical Syntax*. Beth begins by noting (among other things) that, after Frege, '[t]he Kantian conception of pure mathematics was taken up again by intuitionism, as developed by L. Kronecker, H. Poincaré, and, in particular, by L. E. J. Brouwer' (1963, p. 471), and that Carnap, unlike Frege, was explicitly attempting to respond to these developments (among others). The main criticism Beth then develops is that the project of *Logical Syntax* requires what he calls 'a non-formal, intuitive, interpretation' – so that the *Syntax* project is less purely formal, and also less unrestrictedly tolerant, then Carnap appears to realize.

The crux of Beth's argument is that syntax is itself a kind of arithmetic (as becomes especially clear in a Gödel numbering, for example). And, viewed as an arithmetic, a Carnapian syntax language or metalanguage may then have non-standard models – containing non-finite numbers (non-finite sequences of expressions) beyond the standard numbers 0, 1, 2,... (so that, in the case of syntax, there may be more than a finite number of numerals $0, 0^{\prime}, 0^{\prime\prime}, \ldots$, for example, or derivations may have more than a finite number of steps). Someone who understood Carnap's syntax language in accordance with such a non-standard model would systematically misunderstand his main inductive definitions and results; and so, Beth argues, Carnap must implicitly be assuming that the syntax language is understood in accordance with the standard model. Moreover, Beth claims, we are thereby faced with what he calls 'a limitation regarding the Principle of Tolerance' (1963, p. 479); for, although someone who understands Carnap's syntax language in the standard way (such as, presumably, Carnap himself) can understand someone who uses a non-standard interpretation, the perverse practitioner of syntax in accordance with a non-standard interpretation (whom Beth dubs 'Carnap*') would never be able to understand the standard one (for this person lacks precisely the standard understanding of the concept *finite*).

The upshot appears to be, although Beth does not quite say this explicitly, that we must therefore presuppose an intuitive grasp of the standard model of arithmetic prior to the development of any formal calculus or system of logic, and we thereby arrive at a position in the foundations of mathematics more akin to the constructivist or intuitionist tradition (with which Beth is of course much more sympathetic than is Carnap). Moreover, this also means, for Beth, that Carnap's programme cannot avoid all questions of ontological commitment – all questions, for example, about the real existence of numbers outside the context of a particular linguistic framework. For, once again, we must presuppose an intuitive grasp of the standard model of arithmetic prior to the syntactic investigation of any linguistic framework. Thus, Beth concludes, the principle of tolerance 'cannot be accepted without restrictions', and 'Carnap has not been able to avoid every appeal to logical or mathematical intuitions, or, what amounts to the same thing, to ontological commitments' (1963, p. 502).

Carnap, in his reply to Beth, entirely accepts Beth's technical point; and, accordingly, he entirely accepts Beth's claim that '[w]e find in *Logical Syntax* also concepts which, though defined in a purely formal way, are clearly inspired by a non-formal interpretation' (1963b, p. 928). Carnap suggests that he understands this point in terms of the notion of an interpreted formal language or calculus in the sense of the semantical works he developed shortly after *Logical Syntax*. Indeed, in *Foundations of Logic and Mathematics* (1939a), Carnap applies this notion to describe the standard interpretation of Peano arithmetic – where the calculus being interpreted is based on the term '*b*', the functor '...$^{|}$', and the predicate '*N*':

> The *customary interpretation* of the Peano system may first be formulated in this way: '*b*' designates the cardinal number 0; if '...' designates a cardinal number *n*, then '...$^{|}$' designates the next one, i.e., $n+1$; '*N*' designates the class of finite cardinal numbers. Hence in this interpretation the system concerns the progression of finite cardinal numbers, ordered according to magnitude. (1939a/1955, p. 182)

Thus, there is no doubt that Carnap is presupposing the standard understanding of the concept *finite*, just as Beth suggests.

Carnap applies this conception of an interpreted language to the metalanguages used in both syntax and semantics in his reply to Beth:

> Since the metalanguage *ML* serves as a means of communication between the author and the reader or among participants in a discussion, I always presupposed both in syntax and in semantics, that a fixed interpretation

> of *ML*, which is shared by all participants, is given. This interpretation is usually not formulated explicitly; but since *ML* uses English words, it is assumed that these words are understood in their ordinary senses. (1963b, p. 929)

Of course, the imaginary case constructed by Beth violates precisely this presupposition. As Carnap himself puts it: 'Carnap* does not use the metalanguage *ML*, but a language *ML** which, although it uses the same words and sentences, differs from *ML*, since some of the words and sentences have different meanings' (ibid.). Yet Carnap is completely untroubled by this because he is assuming, entirely reasonably, that an unproblematic understanding of the standard model of arithmetic is encapsulated in ordinary mathematical usage. There is no deep mystery here – there is no need to puzzle ourselves over the question how we somehow force an uninterpreted formal calculus to designate or refer to an intended model. We simply give the customary interpretation of this system in unproblematic, antecedently understood terms of ordinary mathematical language; appealing to an 'intuitive grasp' of the standard model adds nothing at all to this routine procedure of ordinary mathematical practice.

Carnap is also completely aware that there are similar cases of failure of communication more directly relevant to his use of the principle of tolerance in *Logical Syntax*:

> It seems to be obvious that, if two men wish to find out whether or not their views on certain objects agree, they must first of all use a common language to make sure that they are talking about the same objects. It may be the case that one of them can express in his own language certain convictions which he cannot translate into the common language; in this case he cannot communicate these convictions to the other man. For example, a classical mathematician is in this situation with respect to an intuitionist or, to a still higher degree, with respect to a nominalist. (1963b, pp. 929–30)

Just as we cannot communicate our standard interpretation of the concept *finite* to Carnap*, the intuitionist cannot understand the classical interpretation of unbounded existential quantification over the natural numbers.

Does this, as Beth suggests, then imply a restriction or limitation of the principle of tolerance? It may at first appear that it does. For Carnap's application of the principle of tolerance here poses the question, in a (syntactic or semantic) metalanguage, whether to adopt the classical or intuitionist logical rules for a particular object-language – in this case, the language of total science (mathematics plus physics). We weigh the relative safety (from the possibility of contradiction) of the intuitionist rules against the greater fruitfulness and convenience (in physics) of the classical rules and then make our choice. But if the intuitionist cannot understand the rules of the classical framework – and cannot, in particular, understand the necessarily even stronger classical metalanguage in which we (syntactically or semantically) describe these rules – then it would appear that our entire procedure simply begs the question against the intuitionist. In no way do we have a neutral shared metaperspective for evaluating the two positions on an equal footing.

This argument is certainly tempting, and I must confess that I myself have succumbed to this temptation more than once (see, for example, Friedman 1999b, ch. 9). I now think, however, that it misses the essence of Carnap's conception, as it emerges particularly clearly in the reply to Beth. Just as in the case of our understanding of the standard model of arithmetic, Carnap presupposes that classical mathematics, as it is standardly practised, is well understood. Indeed, classical mathematics, for Carnap, is a model or paradigm of clear and exact – scientific – understanding, and there is no way, in particular, to raise doubts about our understanding of this framework on the basis of independent purely philosophical commitments. To be sure, the foundations crisis sparked by the discovery of the paradoxes, and the failure of Hilbert's programme, raise serious technical questions relevant to the consistency of the classical framework, and this is precisely why, for Carnap, we should now take intuitionism seriously. To take it seriously, however, means that we entertain the proposal, starting from within the classical framework, that we should weaken its rules to make inconsistency less likely. There is nothing in Carnap's position blocking a classical mathematician from entertaining this option or even deciding then to adopt it. Carnap has therefore not begged the question about the choice between classical and intuitionist mathematics as he understands this question. That an intuitionist mathematician cannot understand the choice as Carnap understands it is irrelevant, for the situation in which we in fact find ourselves has arisen within the paradigmatically well-understood practice of classical mathematics itself.

We noted above that a rejection of pure intuition and the synthetic a priori in favour of the view that all logico-mathematical truth is analytic and has no factual content becomes definitive of what Carnap and the Vienna Circle mean by their empiricism. Indeed, beginning with *Logical Syntax*, Carnap's conception of the analyticity of logico-mathematical truth and his empiricism become correlative and complementary pillars of his overall philosophical position: more specifically, empiricism is understood as the requirement that all *synthetic* sentences of a possible linguistic framework for science should be testable or confirmable via protocol sentences. Our discussion so far has not illuminated this central aspect of Carnap's position, and so it has not yet done justice to the full force of his rejection of pure intuition and the synthetic a priori in favour of tolerance, analyticity, and empiricism. As we have seen, however, at the end of his paper Beth challenges Carnap's views on 'ontological commitment', and this topic, as Carnap indicates in his reply, is of course the main subject of 'Empiricism, Semantics, and Ontology' (1950a). Considering some of the important intervening developments constituting the immediate background to this paper (which remains, to this day, the classic and definitive statement of his overall philosophical position) can further illuminate Carnap's distinctive combination of tolerance, analyticity, and empiricism.

I begin, in particular, with the discussions Carnap held with Tarski and Quine at Harvard during the academic year 1940–1. In these discussions, as we have long known, both Tarski and Quine disputed Carnap's views on analyticity and logical truth (as expressed, at the time, in Carnap's manuscript for his forthcoming *Introduction to Semantics*). However, as examination of Carnap's notes at the Pittsburgh archives by several scholars has recently shown (including Paolo Mancosu [2005] and Greg Frost-Arnold [2006], to both of whom I am indebted here), the main topic of discussion was an attempt to construct a nominalistic version of arithmetic. The idea, as especially promoted by Tarski and Quine, was to develop a nominalistically acceptable conception of mathematics by viewing it as a formal uninterpreted calculus which could nonetheless be used for deductions and proofs via a system of purely syntactic transformation rules. But we know from Gödel (and Carnap) that syntax is essentially arithmetic, so the problem then arises of giving a nominalistically acceptable interpretation of arithmetic itself. Both Tarski and Quine represent the position that full, infinitary classical arithmetic is not meaningful or understandable (*verständlich*) in the strictest sense, and the project they set themselves is to develop a version of finitary arithmetic assuming the existence of nothing other than concrete physical objects – which, for both Tarski and Quine, are paradigmatic of *Verständlichkeit*. Carnap, for his part, does not share at all in these nominalistic philosophical ambitions, but he is interested, as always, in the purely technical problem of seeing how far one can go in the development of calculi or linguistic frameworks subject to a variety of requirements and constraints.

For Quine, the results of these discussions culminated in 'Steps Toward a Constructive Nominalism', published jointly with Nelson Goodman (who had also participated in the earlier discussions) in 1947. This paper begins with the ringing declaration: 'We do not believe in abstract entities' (Goodman and Quine 1947, p. 105); and it goes on to answer the question why the authors 'refuse to admit the abstract objects that mathematics needs' by the statement (ibid.): 'Fundamentally this refusal is based on a philosophical intuition that cannot be justified by anything more ultimate.' Nevertheless, further light is shed on their philosophical motivations by the preceding paragraph:

> Renunciation of abstract objects may leave us with a world composed of physical objects and events, or of units of sense experience, depending upon decisions that need not be made here. Moreover, even when a brand of empiricism is maintained which acknowledges repeatable sensory qualities as well as sensory events, the philosophy of mathematics still faces essentially the same problem that it does when all universals are repudiated. Mere sensory qualities afford no adequate basis for the unlimited universe of numbers, functions, and other classes claimed as values of the variables of classical mathematics. (1947, p. 105)

A footnote to the penultimate sentence then refers us to Goodman's Harvard dissertation, 'A Study of Qualities' (1941), which was largely inspired by Carnap's *Aufbau*, and which eventuated in *The Structure of Appearance* in 1951.

We know that both Goodman and Quine consistently understood the *Aufbau* as a version of empiricist foundationalism in the tradition of Locke, Berkeley, and Hume. Thus Quine, in 'Two Dogmas of Empiricism' (1951), famously considers Carnap's *Aufbau* as the culmination of the 'radical [empiricist] reductionism' developed by Locke and Hume – the doctrine that 'every idea must either originate directly in sense experience or else be compounded out of ideas thus originating' (1951/1953a, p. 38). Carnap reformulated this radical empiricist programme using the formal devices of modern logic, and, in these terms, he almost succeeded: 'He was the first empiricist who, not content with asserting the reducibility of science to terms of immediate experience, took serious steps toward carrying out the reduction' (1951/1953a, p. 39). Similarly, Goodman, in his paper on Carnap's *Aufbau* in the Carnap Schilpp volume, says (1963, p. 558): 'It belongs very much in the main tradition of modern philosophy, and carries forward a little the efforts of the British Empiricists of the 18th Century.' And this suggests that the ultimate philosophical motivations for adopting nominalism, for both Goodman and Quine, derive precisely from the British empiricist tradition.

This suggestion is strikingly confirmed by lectures on Hume's philosophy Quine presented at Harvard in the summer of 1946, which have recently been published in volume 1 of *Eighteenth-Century Thought* (2003). Quine begins with an outline of the history of epistemology very similar in spirit to (although much more detailed than) the sketches he later presents in such celebrated published works as 'Two Dogmas of Empiricism' and 'Epistemology Naturalized' (1969b). Epistemology 'begins as a quest for certainty', 'with the philosophical urge to find a bed-rock of certainty somewhere beneath the probabilities of natural science' (2003, pp. 180–1). This quest culminates, in the modern period, with the rationalism of Descartes and Leibniz based on clear and distinct ideas of reason (paradigmatically exemplified in mathematics) innately implanted in us by God. Fortunately, however,

'Locke made a clean sweep of the whole theory of innate ideas', resulting in the much healthier, and more 'candid', doctrine of empiricism. Here we find the 'bed-rock of certainty' in our 'direct sense impressions', and the programme then becomes one of showing how all '[f]urther ideas are formed from these by combination' (2003, pp. 187–8). We thus arrive at the programme of 'radical [empiricist] reductionism' Quine attributes to Locke and Hume in 'Two Dogmas of Empiricism'.

In his 1946 lectures Quine's discussion of Hume, in particular, takes an especially interesting turn. Quine gives particular emphasis to the fact that 'Hume is a nominalist[; h]e does not believe in universals' (2003, p. 202), and Quine connects this nominalism with Hume's arguments in the *Treatise* that space and time are not infinitely divisible. Quine suggests that a modern version of an 'ideal of empiricist construction' – modelled on Carnap's *Aufbau* but not committed to 'a logic which presupposes universals' – yields the conclusion that 'Hume's condemnation of [geometrical] space remains valid' (2003, p. 209). More precisely, the 'sophisticated', modern construction assumes only propositional connectives, (first-order) quantification, identity, and 'indefinitely many *empirical* predicates'; Hume's questions about infinitely divisible space then become the questions whether all geometrical statements can be expressed in this 'empirically acceptable vocabulary', and whether, so expressed, 'the propositions of infinite divisibility become true' (2003, pp. 209–10). Moreover, 'there is an equal problem, not recognized by Hume, in the infinite divisibility of the numbers themselves – and even in the infinite generability of the whole numbers'. Quine concludes (2003, p. 210): 'In *all* these problems, the answer – even for the sophisticated notion of construction – is very likely *no*.' In sum, from Quine's point of view, Hume has indeed raised a genuine problem about the meaningfulness of classical mathematics (2003, p. 213): '[T]he problem is still alive, and worth reconsidering now from the point of view of an enlightened empiricism – empiricistic and nominalistic as before, but armed with the sophisticated conception of construction.' There can be very little doubt, therefore, that the standards of meaningfulness or *Verständlichkeit* motivating Quine's pursuit of nominalistic arithmetic in the Harvard discussions of 1940–1 – and, quite likely, his work with Goodman in 1947 as well – are precisely those of Humean empiricism (now construed in Quine's 'sophisticated' way).

I noted that Carnap, in the Harvard discussions, does not accept the standards of *Verständlichkeit* appealed to by Tarski and Quine. More generally, he is never attracted to the conception of meaning derived from Lockean and Humean empiricism, according to which only terms directly referring to immediately given concrete sensory data are paradigmatically meaningful. On the contrary, Carnap's conception is quite distinct from traditional empiricism, in that sense experience, on his view, only has significance for science if it is already framed and structured within the abstract forms supplied by logic and mathematics – so that undigested or immediate sense experience, by contrast, is merely private and subjective, with no objective scientific meaning at all. Indeed, this is one of the main themes Carnap develops in the *Aufbau*, and, as late as the Preface to the second edition (1961), Carnap describes his view there as a synthesis of traditional rationalism and traditional empiricism rather than (as in Quine and Goodman) a straightforward and exclusive commitment to traditional empiricism itself:

> For a long time, philosophers of various persuasions have held the view that all concepts and judgments result from the cooperation of experience and reason. In principle, empiricists and rationalists agree in this view, even though the two sides differentially estimate the significance of the two factors, and often obscure the essential agreement by carrying their viewpoints to extremes. The thesis which they have in common is frequently stated in the following simplified version: The senses provide the material of cognition, reason works up [*verarbeitet*] the material into an organized system of cognition. The task thereby arises of establishing a synthesis of traditional empiricism and traditional rationalism. Traditional empiricism rightly emphasized the contribution of the senses, but it did not recognize the significance and peculiarity of logico-mathematical formation. Rationalism had certainly grasped this significance, but it had believed that reason could not only supply form, but could also produce new content out of itself ('a priori'). Through the influence of Gottlob Frege,..., and by studying the works of Bertrand Russell, I had become clear, on the one hand, about the fundamental significance of mathematics for the formation of the system of cognition, and, on the other, about the purely logical, formal character of mathematics on which rests its independence from the contingencies of the real world. These insights formed the basis of my book. (1928a/1967, pp. v–vi; my translation)

Thus, in terms strongly evocative of Kant, Carnap here formulates a version of empiricism which, on the one side, is fundamentally committed to the central role of mathematics in empirical knowledge from the very beginning, and, on the other, also recognizes that this is only possible, in turn, in virtue of its analyticity or complete independence from all factual content.

Carnap provides a particularly clear and extensive discussion of the application of mathematics in empirical science in *Foundations of Logic and Mathematics*. After describing Peano arithmetic and its customary interpretation (§17), Carnap turns to higher (order) mathematical calculi (§18), on the basis of which 'the whole edifice of classical mathematics can be erected without the use of new primitive signs' (1939a/1955, p. 184). The most important fruit of this extension, of course, is the system of analysis of real numbers. Carnap then discusses the application of such mathematical calculi (§19) – with their customary interpretations, of course – and points out that (1939a/1955, p. 186) '[m]athematical calculi with their customary interpretations are distinguished from elementary [first-order] logical calculi chiefly by the occurrence of numerical expressions'. Moreover, '[t]here are two procedures in empirical science which lead to the application of numerical expressions: counting and measurement' (ibid.). The former leads to the application of whole (cardinal) numbers, the latter to the application of the reals. Finally, Carnap provides a detailed analysis of a procedure for testing a very simple physical calculus on the basis of measurements of the three physical magnitudes length, temperature, and thermic expansion (§23). This example makes it especially evident, in particular, that any procedure for testing a physical theory in modern science essentially depends on the quantitative mathematical concepts made available by arithmetic and analysis.

From Carnap's point of view, therefore, there can be absolutely no question of raising empiricist doubts about the meaningfulness of classical mathematics. Carnap's empiricism is based on no prior commitment to any independent philosophical position – and certainly not on Humean empiricism. Empiricism, for Carnap, rather expresses a commitment to the methods of our best empirical physical science, which, once again, constitutes a paradigm, for Carnap, of clear and exact – scientific – understanding. The essential application of classical mathematics (arithmetic and analysis) in this science therefore counts as paradigmatically clear and well understood by the standards of Carnap's empiricism. Indeed, it is in virtue of precisely the same standards that classical mathematics itself counts as clear and well understood. Carnap's response to nominalism, therefore, takes the same form as his response to intuitionism. It certainly makes sense, from the point of view of classical mathematical physical science, to envision a weakening of its logico-mathematical rules: just as the intuitionist can propose to replace Peano arithmetic with the weaker rules of primitive recursive arithmetic, the finitist nominalist can propose to go so far as to weaken the fundamental rules governing the successor function. But it does not make sense to give 'external', purely philosophical reasons for making such proposals: just as it does not make sense, from Carnap's point of view, for the Kantian-inspired intuitionist to question classical unbounded existential quantification on the basis of a prior conception of the necessarily incompletable character of the process of iterating ideal mental operations in pure intuition, it does not make sense for the Hume-inspired nominalist to question the classical rules for successor on the basis of a prior conception of the necessarily particular and concrete character of immediately given sensory data.

I have just used the term 'external' informally in scare-quotes. But the main point of 'Empiricism, Semantics, and Ontology', of course, is to articulate Carnap's attitude toward ontological questions more precisely, using his famous distinction between 'internal' questions, which can be raised and settled within a given linguistic framework introducing this or that type of entities as values of its variables (numbers, physical things, space-time points, and so on), and what Carnap calls 'external questions, i.e., philosophical questions concerning the existence or reality of the total system of the new entities' (1950a/1956, p. 214). With respect to the latter questions, Carnap remarks, '[m]any philosophers regard a question of this kind as an ontological question which must be raised and answered *before* the introduction of the new language forms[; t]he latter introduction, they believe, is legitimate only if it can be justified by an ontological insight supplying an affirmative answer to this question' (ibid.). Carnap's view, on the contrary, is that, although there is certainly a practical question of which such linguistic frameworks to adopt, there is absolutely no corresponding theoretical question (ibid.): 'Above all, it must not be interpreted as referring to an assumption, belief, or assertion of "the reality of the entities"[; t]here is no such assertion[; a]n alleged statement of the reality of the system of entities is a pseudo-statement without cognitive content.'

Towards the end of the paper Carnap then applies this distinction to the case of ontological questions about the existence or reality of numbers, as raised, in particular, by the nominalist:

> The linguistic forms of the framework of numbers, including variables and the general term 'number', are generally used in our common language of communication; and it is easy to formulate explicit rules for their use. Thus the logical characteristics of the framework are sufficiently clear (while many internal questions, i.e., arithmetical questions, are, of course, still open). In spite of this, the controversy concerning the external question of the ontological reality of the system of numbers continues. Suppose that one philosopher says: 'I believe that there are numbers as real entities. This gives me the right to use the linguistic forms of the numerical framework and to make semantical statements about numbers as designata of numerals.' His nominalistic opponent replies: 'You are wrong; there are no numbers. The numerals may still be used as meaningful expressions. But they are not names, there are no entities designated by them. Therefore the word "number" and numerical variables may not be used (unless a way were found of translating them into the nominalistic thing language).' I cannot think of any possible evidence that would be regarded as relevant by both philosophers, and therefore, if actually found, would decide the controversy or at least make one of the opposite theses more probable that the other... Therefore I feel compelled to regard the external question as a pseudo-question, until both parties to the controversy offer a common interpretation of the question as a cognitive question; this would involve an indication of possible evidence regarded as relevant by both sides. (1950a/1956, pp. 218–19)

The position ascribed to the nominalist here corresponds rather closely to that earlier defended by Tarski and Quine at Harvard. And note that Carnap's rejection of the cognitive meaningfulness of this position does not depend on the principle of verifiability, but simply on the idea that any meaningful scientific question must be answerable to one of two types of evidence – logical proof in the case of pure mathematics, empirical evidence in the case of mathematical physical science. Carnap takes it to be obvious that the ontological question of the reality of the system of numbers is not susceptible to mathematical proof; so, if it is cognitively meaningful, it must be answerable to empirical evidence – which, of course, it is not. Thus, once again, it is from the point of view of our best empirical physical science itself that Carnap entirely rejects this question taken as theoretical.

We are now in a position, finally, to see how Carnap's characteristic constellation of commitments to tolerance, analyticity, and empiricism fit together. Empiricism expresses a commitment to our best mathematical physical science as our paradigm of knowledge and conceptual clarity. In this science, pure mathematics – arithmetic and analysis – figures essentially in the very procedure of observational testing that makes it an empirical science in the first place. Absent these branches of mathematics – using only first-order logic and observational vocabulary, for example, with no further commitment to any infinite domain – we have no genuine empirical science at all: we do not have the means for connecting theory to evidence via the well-understood procedures of physical measurement. And this is one important reason for viewing these branches of mathematics as themselves necessarily independent of all contingent empirical matters of fact. But there is another, equally important reason arising from recent disputes in the foundations of mathematics. Carnap puts the point this way in *Foundations of Logic and Mathematics*:

> Concerning mathematics as a pure calculus there are no sharp controversies. These arise as soon as mathematics is dealt with as a system of 'knowledge'; in our terminology, as an interpreted system. Now, if we regard interpreted mathematics as an instrument of deduction within the field of empirical knowledge rather than as a system of information, then many of the controversial problems are recognized as being questions not of truth but of technical expedience. The question is: Which form of the mathematical system is technically most suitable for the purpose mentioned? Which one provides the greatest safety? If we compare, e.g., the systems of classical mathematics and of intuitionistic mathematics, we find that the first is much simpler and technically more efficient, while the second is more safe from surprising occurrences, e.g., contradictions. (1939a/1955, pp. 192–3)

The crucial move, of course, is the conditional assertion expressed in the third sentence. The antecedent of this conditional expresses Carnap's distinctive combination of analyticity and empiricism, the consequent is the principle of tolerance applied to disputes in the foundations of mathematics. It is in precisely this way, therefore, that tolerance, analyticity, and empiricism fit together.

As we have seen, the main dispute in the foundations of mathematics with which Carnap is concerned in 'Empiricism, Semantics, and Ontology' involves nominalism, as recently defended by Tarski, Goodman, and Quine. And it is illuminating to read Carnap's final paragraph with this dispute in mind:

> The acceptance or rejection of abstract linguistic forms, just as the acceptance or rejection of any other linguistic forms in any branch of science, will finally be decided by their efficiency as instruments... To decree dogmatic prohibitions of certain linguistic forms instead of testing them by their success or failure in practical use, is worse than futile; it is positively harmful because it may obstruct scientific progress. The history of science shows examples of such prohibitions based on prejudices deriving from religious, mythological, or other irrational sources, which slowed up the developments for shorter or longer periods of time. Let us learn from the lessons of history.... *Let us be cautious in making assertions and critical in examining them, but tolerant in permitting linguistic forms.* (1950a/1956, p. 221)

Although Carnap is far too diplomatic to say this explicitly, the clear implication, I think, is that the nominalism in question is based on nothing more nor less than an 'irrational' *philosophical* 'prejudice' – a dogmatic commitment to an extreme form of traditional empiricism. Carnap's alternative, at this crucial juncture, needed to guard itself against both intuitionism, on the one side, and nominalistic empiricism, on the other.

# Bibliography

## Unpublished material

ASP = Carnap Nachlass, Archives of Scientific Philosophy, Special Collections, Hillman Library, University of Pittsburgh, Pittsburgh. Quoted by permission of the University of Pittsburgh. All rights reserved.

UCLA = Charles E. Young Research Library, University of California at Los Angeles, Department of Special Collections, Rudolf Carnap papers. Quoted by permission of the Department of Special Collections, Young Research Library, UCLA. All rights reserved.

## Published works

AWODEY Steve, CARUS A. W. (2001) 'Carnap, Completeness, and Categoricity: The *Gabelbarkeitssatz* of 1928', *Erkenntnis* 54, pp. 145–72.

——— (2003) 'Carnap vs. Gödel on Syntax and Tolerance', in P. Parrini, W. Salmon, and M. Salmon (eds), *Logical Empiricism: Historical and Contemporary Perspectives*, Pittsburgh: Pittsburgh University Press, 2003.

——— (2004) 'How Carnap Could Have Replied to Gödel', in Awodey and Klein (2004), pp. 179–200.

——— (2007) 'Carnap's Dream: Gödel, Wittgenstein, and *Logical Syntax*', *Synthese* 159, pp. 23–45.

AWODEY Steve, KLEIN Carsten, eds (2004) *Carnap Brought Home: The View from Jena*, LaSalle, IL: Open Court.

AYER Alfred J., ed. (1959) *Logical Positivism*, Glencoe, IL: Free Press.

BAR-HILLEL Yeoshua (1963) 'Remarks on Carnap's *Logical Syntax of Language*', in Schilpp (1963), pp. 519–43.

BEANEY Michael (2004) 'Carnap's Conception of Explication: From Frege to Husserl?', in Awodey and Klein (2004), pp. 117–50.

BENACERRAF Paul, PUTNAM Hilary, eds (1964) *Philosophy of Mathematics: Selected Readings*, Englewood Cliffs, NJ: Prentice-Hall.

BERGMANN Gustav (1964) *Logic and Reality*, Madison: University of Wisconsin Press, 1964.

BERNAYS Paul (1957) 'Von der Syntax der Sprache zur Philosophie der Wissenschaften', *Dialectica. Revue internationale de philosophie de la connaissance* 11, pp. 233–46.

BETH Evert Willem (1937) 'L'évidence intuitive dans les mathématiques modernes', in *IXe Congrès international de philosophie*, vol. VI, 'Logique et Mathématique', Paris: Hermann, pp. 161–5.

——— (1963) 'Carnap's Views on the Advantages of Constructed Systems Over Natural Languages in the Philosophy of Science', in Schilpp (1963), pp. 468–502.

BOLZANO Bernhard (1837) *Wissenschaftslehre*, Sulzbach, 4 vols. Partially translated by Rolf George as *Theory of Science*, Berkeley, Los Angeles: University of California Press, 1972. Also partially translated by B. Terrell as *Theory of Science*, ed. Jan Berg, Dordrecht: Reidel, 1973.

BONNET Christian, WAGNER Pierre, eds (2006) *L'Âge d'or de l'empirisme logique. Vienne-Berlin-Prague, 1929–1936*, Paris: Gallimard.

CARNAP Rudolf (1922) *Der Raum. Ein Beitrag zur Wissenschaftslehre*, Berlin: Reuther and Reichard.

——— (1927) 'Eigentliche und uneigentliche Begriffe', *Symposion* 1, pp. 355–74.

——— (1928a) *Der logische Aufbau der Welt*, Berlin-Schlachtensee: Weltkreis. 2nd ed., Hamburg: Meiner, 1961. Translated from the 2nd ed. by Rolf George as *The Logical Structure of the World*, Berkeley: University of California Press, 1967; Chicago: Open Court, 2003.

——— (1928b) *Scheinprobleme in der Philosophie: Das Fremdpsychische und der Realismusstreit*, Berlin-Schlachtensee: Weltkreis. Translated by Rolf George as *Pseudoproblems in Philosophy*, Berkeley and Los Angeles: University of California Press, 1967; Chicago: Open Court, 2003.

——— (1929) *Abriss der Logistik, mit besonderer Berücksichtigung der Relationstheorie und ihrer Anwendungen*, Vienna: Julius Springer.

——— (1930a) 'Die alte und die neue Logik', *Erkenntnis* 1, pp. 12–26. Translated by Isaac Levi as 'The Old and New Logic', in Ayer (1959), pp. 133–46.

——— (1930b) 'Die Mathematik als Zweig der Logik', *Blätter für deutsche Philosophie* 4, pp. 298–310.

——— (1930c) '[Review of] Felix Kaufmann, *Das Unendliche in der Mathematik und seine Ausschaltung*', *Deutsche Literaturzeitung* 51, cols. 1674–8.

——— (1930d) 'Bericht über Untersuchungen zur allgemeinen Axiomatik', *Erkenntnis* 1, pp. 303–7.

——— (1931a) 'Die logizistische Grundlegung der Mathematik', *Erkenntnis* 2, pp. 91–105. Translated by Erna Putnam and Gerald J. Massey as 'The Logicist Foundations of Mathematics', in Benacerraf and Putnam (1964), pp. 31–41.

——— (1931b) '[Review of] Kaila, *Der logistische Neupositivismus*, 1930', *Erkenntnis* 2, pp. 75–7.

——— (1931c) '[Beitrag zur] Diskussion zur Grundlegung der Mathematik', *Erkenntnis* 2, pp. 141–4 and pp. 145–6. Translated by John W. Dawson Jr. as '[Contribution to the] Discussion on the Foundation of Mathematics', in Dawson (1984), pp. 120–3 and p. 124.

——— (1932a) 'Überwindung der Metaphysik durch logische Analyse der Sprache', *Erkenntnis* 2, pp. 219–41. Translated by Arthur Pap as 'The Elimination of Metaphysics through Logical Analysis of Language', with 'Remarks by the Author' added in 1957, in Ayer (1959), pp. 60–81.

——— (1932b) 'Die physikalische Sprache als Universalsprache der Wissenschaft', *Erkenntnis* 2, pp. 432–65. Translated by Max Black as Carnap (1934b). Translated into French by Delphine Chapuis-Schmitz as 'La langue de la physique comme langue universelle de la science', in Bonnet and Wagner (2006), pp. 321–62.

——— (1932c) 'Über Protokollsätze', *Erkenntnis* 3, 2/3, pp. 215–28. Translated by Richard Creath and Richard Nollan as 'On Protocol Sentences', *Noûs* 21, 1987, pp. 457–70.

——— (1934a) 'On the Character of Philosophic Problems', translation by W. M. Malisoff, *Philosophy of Science*, 1, pp. 5–19; corrections p. 251. German original published as Carnap (2004b).

——— (1934b) *The Unity of Science*, translation of Carnap (1932b) by Max Black, with a new introduction by Carnap, London: Kegan Paul, Trench, Trubner & Co.

——— (1934c) *Die Aufgabe der Wissenschaftslogik*, Vienna: Gerold. Translated by Hans Kaal as 'The Task of the Logic of Science', in McGuinness (1987), pp. 46–66. Translated into French by Delphine Chapuis-Schmitz, Sandrine Colas, and Pierre Wagner as 'La tâche de la logique de la science', in S. Laugier and P. Wagner, *Philosophie des sciences*, vol. 1, Paris : Vrin, 2004.

———(1934d) *Logische Syntax der Sprache*, Vienna: Springer. Translated as Carnap (1937).

———(1934e) 'Die Antinomien und die Unvollständigkeit der Mathematik', *Monatshefte für Mathematik und Physik* 41, pp. 263–84.

———(1935a) *Philosophy and Logical Syntax*, London: Kegan Paul, Trench, Trubner & Co. Reprint, Bristol: Thoemmes Press, 1996.

———(1935b) 'Formalwissenschaft und Realwissenschaft', *Erkenntnis* 5, pp. 30–7. Translated into French by Pierre Wagner as 'Science formelle et science du réel', in Bonnet and Wagner (2006), pp. 451–9.

———(1935c) 'Les concepts psychologiques et les concepts physiques sont-ils foncièrement différents?', translation by Robert Bouvier, *Revue de Synthèse*, 10, pp. 43–53 [the German original has not been published].

———(1935d) 'Ein Gültigkeitskriterium für die Sätze der klassischen Mathematik', *Monatshefte für Mathematik und Physik* 42, pp. 163–90.

———(1936a) 'Von der Erkenntnistheorie zur Wissenschaftslogik', *Actes du Congrès International de Philosophie Scientifique, Sorbonne, Paris 1935*, fasc. I 'Philosophie scientifique et empirisme logique', Paris: Hermann, pp. 36–41. Translated into French by Pierre Wagner as 'De la théorie de la connaissance à la logique de la science', in Bonnet and Wagner (2006), pp. 519–26.

———(1936b) 'Über die Einheitssprache der Wissenschaft: Logische Bemerkungen zum Projekt einer Enzyklopädie', *Actes du Congrès International de Philosophie Scientifique, Sorbonne, Paris 1935*, fasc. II 'Unité de la science', Paris: Hermann, pp. 60–70.

———(1936c) 'Wahrheit und Bewährung', *Actes du Congrès International de Philosophie Scientifique, Sorbonne, Paris 1935*, fasc. IV, 'Induction et Probabilité', Paris: Hermann, pp. 18–23. Augmented version translated by H. Feigl as 'Truth and Confirmation', in H. Feigl and W. Sellars (eds), *Readings in Philosophical Analysis*, New York: Appleton Century Crofts, 1949, pp. 119–27. Translated into French by Pierre Wagner as 'Vérité et confirmation', in Bonnet and Wagner (2006), pp. 559–72.

———(1936–7) 'Testability and Meaning', *Philosophy of Science* 3, 1936, pp. 419–71 and 4, 1937, pp. 1–40.

———(1937) *The Logical Syntax of Language*, augmented translation of Carnap (1934d) by A. Smeathon, London: Kegan Paul Trench, Trubner & Co and New York: Harcourt, Brace & Co. Reprint, Chicago: Open Court, 2004.

———(1938) 'Logical Foundations of the Unity of Science', in Otto Neurath, Rudolf Carnap, and Charles Morris (eds), *Foundations of the Unity of Science*, vol. 1, Chicago: University of Chicago Press.

———(1939a) *Foundations of Logic and Mathematics, International Encyclopedia of Unified Science*, vol. 1, no. 3, Chicago: University of Chicago Press. Reprinted in Otto Neurath, Rudolf Carnap, and Charles Morris (eds), *Foundations of the Unity of Science*, vol. 1, Chicago: University of Chicago Press, 1955, pp. 139–213.

———(1939b) Review of S. C. Kleene's 'On the Term "Analytic" in Logical Syntax', *Journal of Symbolic Logic* 4, pp. 157–8.

———(1942) *Introduction to Semantics*, Cambridge, MA: Harvard University Press.

———(1945) 'The Two Concepts of Probability', *Philosophy and Phenomenological Research* 5, pp. 513–32.

———(1947) *Meaning and Necessity*, Chicago: University of Chicago Press. 2nd enlarged ed., Chicago and London: University of Chicago Press, 1956.

——— (1950a) 'Empiricism, Semantics, and Ontology', *Revue Internationale de Philosophie* 4, pp. 20–40. Reprinted in Carnap (1956), pp. 205–21.

——— (1950b) *Logical Foundations of Probability*, Chicago: University of Chicago Press; 2nd ed., 1962.

——— (1954) *Einführung in die symbolische Logik, mit besonderer Berücksichtigung ihrer Anwendungen*, Vienna: Springer. Translated by W. Meyer and J. Wilkinson as *Introduction to Symbolic Logic and its Applications*, New York: Dover, 1958.

——— (1955) 'Meaning and Synonymy in Natural Languages', *Philosophical Studies* 6 (3), pp. 33–47. Reprinted in Carnap (1956), pp. 233–47.

——— (1956) *Meaning and Necessity*, 2nd ed., Chicago: University of Chicago Press.

——— (1962) *Logical Foundations of Probability*, 2nd ed., Chicago: University of Chicago Press.

——— (1963a) 'Intellectual Autobiography', in Schilpp (1963), pp. 3–84.

——— (1963b) 'Replies and Systematic Expositions', in Schilpp (1963), pp. 859–1013.

——— (2000) *Untersuchungen zur allgemeinen Axiomatik*, ed. T. Bonk and J. Mosterin, Darmstadt: Wissenschaftliche Buchgesellschaft.

——— (2004a) *Scheinprobleme in der Philosophie und andere metaphysikkritische Schriften*, ed. Thomas Mormann, Hamburg: Felix Meiner.

——— (2004b) 'Über den Charakter der philosophischen Probleme', in Carnap (2004a), pp. 111–27. Translated by W. M. Malisoff as Carnap (1934a).

——— (forthcoming) *The Collected Works of Rudolf Carnap*, Chicago: Open Court.

[CARNAP Rudolf, HAHN Hans, NEURATH Otto]* (1929) *Wissenschaftliche Weltauffassung. Der Wiener Kreis*, Vienna: Gerold. Translated by Paul Foulkes and Marie Neurath as 'The Scientific World Conception: The Vienna Circle', in Otto Neurath, *Empiricism and Sociology*, ed. M. Neurath and R. S. Cohen, Reidel: Dordrecht, 1973, pp. 299–318.

CARUS A. W. (1999) 'Carnap, Syntax, and Truth', in Jaroslav Peregrin (ed.), *Truth and its Nature (If Any)*, Dordrecht: Kluwer, pp. 15–35.

——— (2004) 'Sellars, Carnap, and the Logical Space of Reasons', in Awodey and Klein (2004), pp. 317–55.

——— (2007) *Carnap and Twentieth-Century Thought: Explication as Enlightenment*, Cambridge: Cambridge University Press.

CHAPUIS-SCHMITZ Delphine (2006) 'Le sens à l'épreuve de l'expérience. Carnap, Schlick et le vérificationnisme', PhD Thesis, University Paris 1 Panthéon-Sorbonne.

CHURCH Alonzo (1932) 'A Set of Postulates for the Foundation of Logic', *The Annals of Mathematics*, 2nd series, 33, pp. 346–66.

——— (1956) *Introduction to Mathematical Logic*, Princeton: Princeton University Press.

COFFA Alberto (1987) 'Carnap, Tarski, and the Search for Truth', *Noûs* 21, pp. 547–72.

——— (1991) *The Semantic Tradition from Kant to Carnap: To the Vienna Station*, Cambridge: Cambridge University Press.

CONANT James (2001) 'Two Conceptions of *Die Überwindung der Metaphysik*: Carnap and Early Wittgenstein', in T. G. McCarthy and S. C. Stidd (eds), *Wittgenstein in America*, Oxford: Oxford University Press, pp. 13–61.

CREATH Richard (1987) 'The Initial Reception of Carnap's Doctrine of Analyticity', *Noûs* 21 (4), pp. 477–99.

——— ed. (1990) *Dear Carnap, Dear Van: The Quine–Carnap Correspondence and Related Work*, Los Angeles: University of California Press.

* These brackets mean that this text (known as the 'Vienna Circle Manifesto') was published anonymously (although its short preface was signed).

——— (1991) 'The Unimportance of Semantics', *PSA1990: Proceedings of the Biennial Meeting of the Philosophy of Science Association*, vol. 2, pp. 405–16.
——— (1992) 'Carnap's Conventionalism', *Synthese* 93, pp. 141–65.
——— (1996) 'Languages Without Logic', in Giere and Richardson (1996), pp. 251–68.
——— (1999) 'Carnap's Move to Semantics: Gains and Losses', in Jan Wolenski and Eckehart Köhler (eds), *Alfred Tarski and the Vienna Circle*, Dordrecht: Kluwer, 1999, pp. 65–76.
——— (2004) 'Carnap's Program and Quine's Question', in Awodey and Klein (2004), pp. 279–93.
——— (forthcoming) 'The Logical and the Analytic'.
DAVIDSON Donald, HINTIKKA Jaakko, eds (1969) *Words and Objections: Essays on the Work of W. V. Quine*, Dordrecht: D. Reidel.
DAWSON Jr. John W. (1984) 'Discussion on the Foundations of Mathematics', *History and Philosophy of Logic* 5, pp. 111–29.
DREBEN Burton (1990) 'Quine', in Robert Barrett and Roger Gibson (eds), *Perspectives on Quine*, Cambridge, MA: Basil Blackwell, pp. 81–95.
DUMMETT Michael (1981) *Frege: Philosophy of Language*, 2nd ed., London: Duckworth.
EINSTEIN Albert (1921) *Geometrie und Erfahrung*, Berlin: Springer.
ENDERTON Herbert B. (1972) *A Mathematical Introduction to Logic*, San Diego: Academic Press, 2nd ed. 2001.
FLOYD Juliet (2002) 'Number and Ascriptions of Number in Wittgenstein's *Tractatus*', in E. Reck (ed.), *From Frege to Wittgenstein: Perspectives on Early Analytic Philosophy*, Oxford: Oxford University Press, pp. 308–52.
FRAENKEL Abraham (1928) *Einleitung in der Mengenlehre*, 3rd ed., Berlin: Springer.
FREGE Gottlob (1980) *Philosophical and Mathematical Correspondence*, ed. Gottfried Gabriel et al., translated by Hans Kaal, Chicago: University of Chicago Press, 1980.
FRIEDMAN Michael (1988) 'Logical Truth and Analyticity in Carnap's "Logical Syntax of Language"', in W. Aspray and P. Kitcher (eds), *History and Philosophy of Modern Mathematics* (Minnesota Studies in the Philosophy of Science, vol. XI), Minneapolis: University of Minnesota Press, pp. 82–94. Reprinted as 'Analytic Truth in Carnap's *Logical Syntax of Language*', in Friedman (1999b), pp. 165–76.
——— (1991) 'The Re-evaluation of Logical Positivism', *Journal of Philosophy* 88, pp. 505–19. Reprinted in Friedman (1999b), pp. 1–14.
——— (1994) 'Geometry, Convention and the Relativized A Priori: Reichenbach, Schlick and Carnap', in W. Salmon and G. Wolters (eds), *Logic, Language and the Structure of Scientific Theories*, Pittsburgh: University of Pittsburgh Press, 1994, pp. 21–34. Reprinted in Friedman (1999b), pp. 59–70.
——— (1997) 'Carnap and Wittgenstein's *Tractatus*', in W. W. Tait (ed.), *Early Analytical Philosophy*, La Salle, IL: Open Court, pp. 19–36. Reprinted in Friedman (1999b), pp. 177–97.
——— (1999a) 'Tolerance and Analyticity in Carnap's Philosophy of Mathematics', in Friedman (1999b), pp. 198–233. Reprinted in J. Floyd and S. Shieh (eds), *Future Pasts: The Analytic Tradition in Twentieth-Century Philosophy*, Oxford: Oxford University Press, 2001, pp. 223–55.
——— (1999b) *Reconsidering Logical Positivism*, Cambridge: Cambridge University Press.
——— (2000) *A Parting of the Ways: Carnap, Cassirer, Heidegger*, LaSalle, IL: Open Court.
——— (2001) *The Dynamics of Reason*, Stanford: CSLI.
——— (2007) 'The *Aufbau* and the Rejection of Metaphysics', in Friedman and Creath (2007), pp. 129–52.

——— (2008) 'Carnap and Quine: Twentieth-Century Echoes of Kant and Hume', *Philosophical Topics* 34.

FRIEDMAN Michael, CREATH Richard, eds (2007) *The Cambridge Companion to Carnap*, Cambridge: Cambridge University Press.

FROST-ARNOLD Gregg (2006) 'Carnap, Tarski, and Quine's Year Together: Logic, Science, and Mathematics', Doctoral Dissertation, University of Pittsburgh.

GEACH Peter (1983) 'Wittgenstein's Operator *N*', *Analysis* 41, pp. 573–89.

GIERE Ronald N., RICHARDSON Alan W., eds (1996) *Origins of Logical Empiricism* (Minnesota Studies in the Philosophy of Science, vol. XVI), Minneapolis: University of Minnesota Press.

GILLIES Donald (1981) 'Karl Menger as a Philosopher', *British Journal for the Philosophy of Science* 32, pp. 183–96.

GÖDEL Kurt (1931) 'Über formal unentscheidbare Sätze der Principia Mathematica und verwandter Systeme I', *Monatshefte für Mathematik und Physik* 38, pp. 173–98. Reprinted in Gödel, *Collected Works*, vol. I: *Publications 1929–1936*, ed. S. Feferman et al., Oxford: Oxford University Press, 1986.

——— (1934) 'On Undecidable Propositions of Formal Mathematical Systems', mimeographed lecture notes, taken by S. C. Kleene and J. B. Rosser, reprinted with revisions in Martin Davis (ed.), *The Undecidable*, Hewlett, NY: Raven Press, 1965, pp. 41–74. Reprinted in Gödel, *Collected Works*, vol. 1, 1986, pp. 346–71.

——— (1995) 'Is Mathematics Syntax of Language?', version III and version V, in Gödel, *Collected Works*, vol. III: *Unpublished Essays and Lectures*, ed. S. Feferman et al., Oxford: Oxford University Press, pp. 334–62.

——— (2003) *Collected Works*, vol. IV: *Correspondence A–G*, Oxford: Oxford University Press.

GOLDFARB Warren (1989) 'Russell's Reasons for Ramification', in C. W. Savage and C. A. Anderson (eds), *Rereading Russell: Essays on Bertrand Russell's Metaphysics and Epistemology* (Minnesota Studies in the Philosophy of Science, vol. XII), Minneapolis: University of Minnesota Press, pp. 24–40.

——— (1995) 'Introductory Note to *1953/9', in Gödel (1995), pp. 324–34.

——— (1996) 'The Philosophy of Mathematics in Early Positivism', in Giere and Richardson (1996), pp. 213–30.

——— (1997) 'Semantics in Carnap: A Rejoinder to Alberto Coffa', *Philosophical Topics* 25, pp. 51–66.

——— (2003) 'Introductory Note', in Gödel (2003), pp. 335–42.

——— (2005) 'On Gödel's Way In: The Influence of Rudolf Carnap', *Bulletin of Symbolic Logic* 11/2, pp. 185–93.

GOLDFARB Warren, RICKETTS Thomas (1992) 'Carnap and the Philosophy of Mathematics', in D. Bell and W. Vossenkuhl (eds), *Science and Subjectivity: The Vienna Circle and Twentieth Century Philosophy*, Berlin: Akademie-Verlag, pp. 61–78.

GOODMAN Nelson (1951) *The Structure of Appearance*, Cambridge, MA: Harvard University Press.

——— (1963) 'The Significance of *Der logische Aufbau der Welt*', in Schilpp (1963), pp. 545–58.

GOODMAN Nelson, QUINE Willard Van Orman (1947) 'Steps Toward a Constructive Nominalism', *Journal of Symbolic Logic* 12, pp. 105–22.

HACKER Peter M. S. (2003) 'Wittgenstein, Carnap, and the New American Wittgensteinians', *Philosophical Quarterly* 53, pp. 1–23.

HAHN Hans (1929) 'Empirismus, Mathematik, Logik', *Forschungen und Fortschritte* 5, pp. 409–10. Reprinted in Hahn (1988), pp. 55–8. Translated by Hans Kaal as 'Empiricism, Mathematics, and Logic', in Hahn (1980), pp. 39–42.

——— (1930a) 'Die Bedeutung der wissenschaftlichen Weltauffassung insbesondere für Mathematik und Physik', *Erkenntnis* 1, pp. 96–105. Reprinted in Hahn (1988). Translated by Hans Kaal as 'The Significance of the Scientific World View, Especially for Mathematics and Physics', in Hahn (1980), pp. 20–30.

——— (1930b) *Überflussige Wesenheiten (Occams Rasiermesser)*, Vienna: Wolf. Reprinted in Hahn (1988), pp. 21–37. Translated by Hans Kaal as 'Superfluous Entities, or Occam's Razor', in Hahn (1980), pp. 1–19.

——— (1931) '[Beitrag zur] Diskussion zur Grundlegung der Mathematik', *Erkenntnis* 2, pp. 135–41. Reprinted in Hahn (1988). Translated by Hans Kaal as 'Discussion about the Foundations of Mathematics', in Hahn (1980), pp. 31–8 and by John Dawson Jr., in Dawson (1984), pp. 111–17.

——— (1933a) *Logik, Mathematik, Naturerkennen*, Vienna: Gerold. Reprinted in Hahn (1988), pp. 141–72. Translated by Hans Kaal as 'Logic, Mathematics and Knowledge of Nature', in McGuinness (1987), pp. 24–45.

——— (1933b) 'Die Krise der Anschauung', in *Krise und Neuaufbau in der Exakten Wissenschaften. Fünf Wiener Vorträge*, Vienna: Deuticke, pp. 41–64. Reprinted in Hahn (1988), pp. 86–114. Translated as 'The Crisis in Intuition', in Hahn (1980), pp. 73–102.

——— (1934) 'Gibt es Unendliches?', in *Alte Probleme – Neue Lösungen in den exakten Wissenschaften. Fünf Wiener Vorträge*, Vienna: Deuticke, pp. 93–116. Reprinted in Hahn (1988). Translated as 'Does the Infinite Exist?', in Hahn (1980), pp. 103–31.

——— (1980) *Empiricism, Logic and Mathematics*, ed. B. McGuinness, Dordrecht: Reidel.

——— (1988) *Empirismus, Logik, Mathematik*, Frankfurt am Main: Suhrkamp.

HAHN Hans et al. (1931) 'Diskussion zur Grundlegung der Mathematik', *Erkenntnis* 2, pp. 135–49. Translated by John Dawson Jr. as Dawson (1984).

HEIJENOORT Jean van, ed. (1967) *From Frege to Gödel: A Source Book in Mathematical Logic. 1879–1931*, Cambridge, MA: Harvard University Press.

HILBERT David, ACKERMANN Wilhelm (1928) *Grundzüge der theoretischen Logik*, Berlin: Springer.

JEFFREY Richard (1994) 'Carnap's Voluntarism', in D. Prawitz, B. Skyrms, and D. Westerståhl (eds), *Logic, Methodology, and Philosophy of Science IX*, Amsterdam: Elsevier, pp. 847–66.

KAILA Eino S. (1930) *Der logistische Neupositivismus. Eine kritische Studie*, Turku, *Turun Yliopiston Jukaisuja. Annales Universitatis Aboensis*, Series B, vol. XIII. Translated by Ann and Peter Kirschenmann as 'Logistic Neopositivism: A Critical Study', in Kaila, *Reality and Experience*, ed. R. S. Cohen, Dordrecht: Reidel, 1979, pp. 1–58.

KAUFMANN Felix (1930) *Das Unendliche in der Mathematik und seine Ausschaltung*, Leipzig: Deuticke. Translated by Paul Foulkes as 'The Infinite in Mathematics and its Elimination', in Kaufmann, *The Infinite in Mathematics*, ed. B. McGuinness, Dordrecht: Kluwer, 1978, pp. 1–164.

KIENZLER Wolfgang (2008) 'Wittgenstein und Carnap: Klarkeit oder Deutlichkeit als Ideal der Philosophie', in Christiane Schildknecht, Dieter Teichert, Temilo van Zantwijk (eds), *Genese und Geltung. Für Gottfried Gabriel*, Paderborn: Mentis.

KLEENE Stephen C. (1939a) 'Review of Rudolf Carnap, *The Logical Syntax of Language*', *Journal of Symbolic Logic* 4/2, pp. 82–7.

——— (1939b) 'On the Term "Analytic" in Logical Syntax', *Journal of Unified Science (Erkenntnis)*, 9 (the volume, which was ready for publication, remained unpublished), pp. 189–92.

KLEIN Felix (1872) 'Vergleichende Betrachtungen über neuere geometrische Forschungen', *Mathematische Annalen* 43, 1893, pp. 63–100 (lecture read in Erlangen in Oct. 1872).

KOELLNER Peter (forthcoming) 'Truth in Mathematics: The Question of Pluralism'.

KÖHLER Eckehart (1991) 'Gödel und der Wiener Kreis', in P. Kruntorad (ed.), *Jour Fixe der Vernunft*, Vienna: Holder-Pichler-Tempsky, pp. 127–64.

——— (1993) 'Gödel und Carnap in Wien und Prag', in R. Haller and F. Stadler (eds), *Wien-Berlin-Prag. Der Aufstieg der wissenschaftlichen Philosophie*, Vienna: Holder-Pichler-Tempsky, pp. 165–74.

KOKOSZYŃSKA Maria (1936) 'Über den absoluten Wahrheitsbegriff und einige andere semantische Begriffe', *Erkenntnis* 6, pp. 143–65.

LAVERS Gregory (2004) 'Carnap, Semantics, and Ontology', *Erkenntnis* 60, pp. 295–316.

MAC LANE Saunders (1938) 'Carnap on Logical Syntax', *Bulletin of the American Mathematical Society* 44, pp. 171–6.

MANCOSU Paolo (2002) 'On the Constructivity of Proofs: A Debate among Behmann, Bernays, Gödel, and Kaufmann', in W. Sieg, R. Sommer, and C. Talcott (eds), *Reflections on the Foundations of Mathematics. Essays in Honor of Solomon Feferman*, Association for Symbolic Logic (Lecture Notes in Logic, vol. 15), pp. 346–68.

——— (2005) 'Harvard 1940–41: Tarski, Carnap and Quine on a Finitist Language for Mathematics and Science', *History and Philosophy of Logic* 26, pp. 327–57.

——— (2008a) 'Quine and Tarski on Nominalism', in *Oxford Studies in Metaphysics*, vol. IV, pp. 22–55.

——— (2008b) 'Tarski, Neurath, and Kokoszyńska on the Semantic Conception of Truth', in Douglas Patterson (ed.), *New Essays on Tarski and Philosophy*, New York: Oxford University Press, pp. 192–224.

MCGUINNESS Brian, ed. (1967) *Wittgenstein und der Wiener Kreis. Gespräche aufgezeichnet von Friedrich Waismann*, Frankfurt am Main: Suhrkamp. Translated by J. Schulte and B. McGuiness as *Wittgenstein and the Vienna Circle: Conversations Recorded by Friedrich Waismann*, Oxford: Blackwell, 1979.

——— ed. (1987) *Unified Science*, Dordrecht: Reidel.

MENGER Karl (1928) 'Bemerkungen zu Grundlagenfragen I', *Jahrbuch der Deutschen Mathematikervereinigung* 37, pp. 213–26. Translated by the author as 'An Intuitionistic-Formalistic Dictionary of Set Theory', in Menger (1979b), pp. 79–87.

——— (1930) 'Der Intuitionismus', *Blätter für Deutsche Philosophie* 4, pp. 311–25. Translated by Robert Kowalski as 'On Intuitionism' in Menger (1979b), pp. 46–58.

——— (1933) 'Die neue Logik', in *Krise und Neuaufbau in der exakten Wissenschaften. Fünf Wiener Vorträge*, Vienna: Deuticke, pp. 94–122. Translation of a revised edition by the author: 'The New Logic', *Philosophy of Science* 4, 1937, pp. 299–336. Translation reprinted in Menger (1979b), pp. 17–45.

——— (1979a) 'Logical Tolerance in the Vienna Circle', in Menger (1979b), pp. 11–16.

——— (1979b) *Selected Papers in Logic and Foundations, Didactics, Economics*, Dordrecht: Reidel.

——— (1982) 'Memories of Schlick', in E. T. Gadol (ed.), *Rationality and Science*, Vienna: Springer, pp. 83–103.

——— (1994) *Reminiscences of the Vienna Circle and the Mathematical Colloquium*, Dordrecht: Kluwer.

METSCHL Ulrich (1992) 'Toleranz und Pluralismus. Die *Logische Syntax der Sprache* und die Behandlung logischer Konstanten', in D. Bell and W. Vossenkuhl (eds), *Science and Subjectivity: The Vienna Circle and Twentieth Century Philosophy*, Berlin: Akademie-Verlag, pp. 79–99.

NORTON Bryan (1977) *Linguistic Frameworks and Ontology: A Re-examination of Carnap's Metaphilosophy*, The Hague, New York and Paris: Mouton.

OBERDAN Thomas (1992) 'The Concept of Truth in Carnap's *Logical Syntax of Language*', *Synthese* 93, pp. 239–60.

——— (1993) *Protocols, Truth and Convention*, Amsterdam: Rodopi.

——— (1996) 'Postscript to Protocols: Reflections on Empiricism', in Giere and Richardson (1996), pp. 269–87.

——— (2004) 'Carnap's Conventionalism: The Problem with P-Rules', *Grazer Philosophische Studien* 68, pp. 119–38.

O'GRADY Paul (1999) 'Carnap and Two Dogmas of Empiricism', *Philosophy and Phenomenological Research* 59 (4), pp. 1015–27.

POPPER Karl (1963) 'The Demarcation Between Science and Metaphysics', in Schilpp (1963), pp. 183–226.

POTTER Michael D. (2000) *Reason's Nearest Kin: Philosophies of Arithmetic from Kant to Carnap*, Oxford and New York: Oxford University Press.

PROOPS Ian (2001) 'The New Wittgenstein: A Critique', *European Journal of Philosophy* 9, pp. 375–404.

PUTNAM Hilary W. (1981) *Reason, Truth, and History*, Cambridge: Cambridge University Press.

QUINE Willard van Orman (1936) 'Truth by Convention', in O. H. Lee (ed.), *Philosophical Essays for A. N. Whitehead*, New York: Longmans. Reprinted in Quine (1976), pp. 77–106.

——— (1951) 'Two Dogmas of Empiricism', *Philosophical Review* 60, pp. 20–43. Reprinted in Quine (1953a/1961), pp. 20–46.

——— (1953a) *From a Logical Point of View*, New York: Harper, 2nd ed., Cambridge, MA: Harvard University Press, 1961.

——— (1953b) 'The Problem of Meaning in Linguistics', in Quine (1953a/1961), pp. 47–64.

——— (1953c) 'Mr. Strawson on Logical Theory', *Mind* 62, pp. 433–51. Reprinted in Quine (1976), pp. 137–57.

——— (1957) 'The Scope and Language of Science', *British Journal for the Philosophy of Science* 8, pp. 1–17. Reprinted in Quine (1976), pp. 228–45.

——— (1958) 'Speaking of Objects', *Proceedings and Addresses of the American Philosophical Association*, vol. 31, pp. 5–22. Reprinted in Quine (1969a), pp. 1–25.

——— (1960) *Word and Object*, Cambridge, MA: MIT Press.

——— (1963) 'Carnap and Logical Truth', in Schilpp (1963), pp. 385–406. Reprinted in Quine (1976), pp. 107–32. Originally in *Synthese* 12, 1960, pp. 350–74.

——— (1969a) *Ontological Relativity and Other Essays*, New York: Columbia University Press.

——— (1969b) 'Epistemology Naturalized', in Quine (1969a), pp. 69–90.

——— (1969c) 'Reply to Strawson', in Davidson and Hintikka (1969), pp. 320–25.

——— (1970) *Philosophy of Logic*, Cambridge, MA: Harvard University Press; 2nd ed., 1986.

——— (1973) *Roots of Reference*, La Salle: Open Court.

——— (1976) *The Ways of Paradox and Other Essays*, 2nd ed., revised and enlarged, Cambridge, MA: Harvard University Press.

——— (1981a) *Theories and Things*, Cambridge, MA: Harvard University Press.

——— (1981b) 'Things and Their Place in Theories', in Quine (1981a), pp. 1–23.

——— (1986) *Philosophy of Logic*, 2nd ed., Cambridge, MA: Harvard University Press.

——— (1990a) 'Letter to Carnap, February 4, 1938', in Creath (1990), pp. 239–44.

———(1990b) *Pursuit of Truth*, Cambridge, MA: Harvard University Press. Revised ed., 1992.
———(1993) 'In Praise of Observational Sentences', *Journal of Philosophy* 90, pp. 107–16.
———(2003) '1946 Lectures on David Hume's Philosophy', ed. James G. Buickerrood, *Eighteenth-Century Thought* 1, pp. 171–254.
RECK Erich H. (2007) 'Carnap and Modern Logic', in Friedman and Creath (2007), pp. 176–99.
REICHENBACH Hans (1929) 'Ziele und Wege der physikalischen Erkenntnis', in *Handbuch der Physik*, vol. 4: *Allgemeine Grundlagen der Physik*, pp. 1–89, Berlin: Springer.
RICHARDSON Alan W. (1994) 'The Limits of Tolerance: Carnap's Logico-Philosophical Project in *Logical Syntax of Language*', *Proceedings of the Aristotelian Society*, Supplementary Volume 67, pp. 67–82.
———(1996) 'From Epistemology to the Logic of Science: Carnap's Philosophy of Empirical Knowledge in the 1930s', in Giere and Richardson (1996), pp. 309–32.
———(1997) 'Two Dogmas about Logical Empiricism: Carnap and Quine on Logic, Epistemology, and Empiricism', *Philosophical Topics* 25, pp. 145–68.
———(1998) *Carnap's Construction of the World: The* Aufbau *and the Emergence of Logical Empiricism*, Cambridge and New York: Cambridge University Press.
RICKETTS Thomas (1994) 'Carnap's Principle of Tolerance, Empiricism, and Conventionalism', in P. Clark and B. Hale (eds), *Reading Putnam*, Oxford: Blackwell, pp. 176–200.
———(1996) 'Carnap: From Logical Syntax to Semantics', in Giere and Richardson (1996), pp. 231–50.
———(2003) 'Languages and Calculi', in G. Hardcastle and A. Richardson (eds), *Logical Empiricism in North America* (Minnesota Studies in the Philosophy of Science, vol. XVIII), Minneapolis: University of Minnesota Press, 2003, pp. 257–80.
———(2004) 'Frege, Carnap, and Quine: Continuities and Discontinuities', in Awodey and Klein (2004), pp. 181–202.
———(2007) 'Tolerance and Logicism: Logical Syntax and the Philosophy of Mathematics', in Friedman and Creath (2007), pp. 200–25.
ROSSER J. Barkley (1936) 'Extensions of Some Theorems of Gödel and Church', *Journal of Symbolic Logic* 1, pp. 87–91.
———(1937) 'Gödel Theorems for Non-Constructive Logics', *Journal of Symbolic Logic* 2, pp. 129–37.
ROUILHAN Philippe de (1998–9) 'Les tableaux de Beth: syntaxe ou sémantique?', *Philosophia Scientiae* 3, pp. 302–22.
———(2008) 'Sur la conséquence logique pour les Langages I et II de la *Syntaxe* de Carnap', unabridged, French version of 'Carnap on Logical Consequence for Languages I and II' (this volume), http://www-ihpst.univ-paris1.fr.
RUSSELL Bertrand (1914) *Our Knowledge of the External World as a Field for Scientific Method in Philosophy*, Chicago: Open Court.
———(1919) *Introduction to Mathematical Philosophy*, London: Allen & Unwin.
———(1922) 'Introduction' to Wittgenstein's *Tractatus*. Reprinted in Wittgenstein *Logisch-philosophische Abhandlung*, ed. B. McGuinness and J. Schulte, Frankfurt am Main: Suhrkamp, 1988, pp. 258–86.
———(1959) *My Philosophical Development*, London: Allen & Unwin.
RUSSELL Bertrand, WHITEHEAD Alfred North (1910–13), *Principia Mathematica*, vols 1–3, Cambridge: Cambridge University Press.

RYCKMAN Thomas A. (1992) 'P(oint) C(oincidence) Thinking: The Ironical Attachment of Logical Empiricism to General Relativity (And Some Lingering Consequences)', *Studies in History and Philosophy of Science* 23, pp. 471–93.

——— (2005) *The Reign of Relativity: Philosophy in Physics 1915–1925*, Oxford: Oxford University Press.

SARKAR Sahotra (1992) ' "The Boundless Ocean of Unlimited Possibilities": Logic in Carnap's *Logical Syntax of Language*', *Synthese* 93, pp. 191–237.

SCHILPP Paul, ed. (1963) *The Philosophy of Rudolf Carnap*, Chicago and La Salle: Open Court.

SCHLICK Moritz (1925) *Allgemeine Erkenntnislehre*, 2nd ed., Berlin: Julius Springer. Reprint Frankfurt am Main: Suhrkamp, 1979. Translated by Albert E. Blumberg as *General Theory of Knowledge*, New York: Springer, 1974.

——— (1930) 'Die Wende der Philosophie', *Erkenntnis* 1, pp. 4–11. Translated by David Rynin as 'The Turning Point in Philosophy', in Ayer (1959), pp. 53–9. Translated into French by Delphine Chapuis-Schmitz as 'Le tournant de la philosophie', in Laugier and Wagner (eds), *Philosophie des sciences*, vol. 1, Paris: Vrin, 2004, pp. 177–86.

——— (1931) 'Die Kausalität in der gegenwärtigen Physik', *Die Naturwissenschaften* 19, 145–62. Translated by Peter Heath as 'Causality in Contemporary Physics', in Schlick (1979), pp. 176–209. Translated into French by Céline Vautrin as 'La causalité dans la physique contemporaine', in Bonnet and Wagner (2006), pp. 171–219.

——— (1934) 'Über das Fundament der Erkenntnis', *Erkenntnis* 4, pp. 79–99. Translated into French as Schlick (1935a). Translated by Peter Heath as 'On the Foundation of Knowledge', in Schlick (1979), pp. 370–87. Translated into French by Delphine Chapuis-Schmitz as 'Sur le fondement de la connaissance', in Bonnet and Wagner (2006), pp. 415–39.

——— (1935a) 'Sur le fondement de la connaissance', French translation of Schlick (1934) by Ernest Vouillemin, in Schlick (1935c), pp. 8–34.

——— (1935b) 'Sur les "Constatations"', in Schlick (1935c) pp. 44–54. Translated by Peter Heath as 'On "Affirmations"', in Schlick (1979), pp. 407–13.

——— (1935c) *Sur le fondement de la connaissance*, Paris: Hermann.

——— (1936) 'Sind die Naturgesetze Konventionen?', in *Actes du Congrès International de Philosophie Scientifique, Paris 1935*, fasc. IV 'Induction et Probabilité', Paris: Hermann, pp. 8–17. Translated by Herbert Feigl and May Brodbeck as 'Are Natural Laws Conventions?', in Schlick (1979), pp. 437–45. Translated into French by Céline Vautrin as 'Les lois de la nature sont-elles des conventions?', in Bonnet and Wagner (2006), pp. 537–48.

——— (1979) *Philosophical Papers*, vol. 2, ed. H. L. Mulder and B. van de Velde-Schlick, Dordrecht: Reidel.

SCHURZ Gerhard (1999) 'Tarski and Carnap on Logical Truth – or: What is Genuine Logic?', in Jan Woleński and Eckehart Köhler (eds), *Alfred Tarski and the Vienna Circle*, Dordrecht: Kluwer, 1999, pp. 77–94.

SKOLEM Thoralf (1923) 'Begründung der elementaren Arithmetik durch die rekurrierende Denkeweise ohne Anwendung scheinbarer Veränderlichen mit unendlichem Ausdehnungsbereich', *Videnskapsselskapets skrifter, I. Matematisk-naturvidenskabelig klasse*, no. 6. Translated by S. B. Bauer-Mengelberg as 'The Foundations of Elementary Arithmetic by Means of Recursive Mode of Thought without the Use of Apparent Variables Ranging over Infinite Domains', in Heijenoort (1967), pp. 302–33.

SOAMES Scott (1983) 'Generality, Truth Functions, and Expressive Capacity', *Philosophical Review* 92, pp. 573–89.

STADLER Friedrich (1997) *Studien zum Wiener Kreis: Ursprung, Entwicklung und Wirkung des logischen Empirismus im Kontext,* Frankfurt am Main: Surkamp. Translated by C. Nielsen, J. Golb, S. Schmidt, and T. Ernst as *The Vienna Circle: Studies in the Origins, Development and Influence of Logical Empiricism,* Vienna and New York: Springer, 2001.

STEIN Howard (1992) 'Was Carnap Entirely Wrong, After All?', *Synthese* 93, pp. 275–95.

STERN David (2007) 'Wittgenstein, the Vienna Circle and Physicalism: A Reassessment', in A. Richardson and T. Uebel (eds), *Cambridge Companion to Logical Empiricism,* Cambridge: Cambridge University Press.

STRAWSON Peter Frederick (1961) 'Singular Terms and Predication', *Journal of Philosophy* 58, pp. 393–412. Reprinted in Davidson and Hintikka (1969), pp. 97–117.

SUNDHOLM Göran (1992) 'The General Form of the Operation in Wittgenstein's *Tractatus*', *Grazer Philosophische Studien* 42, pp. 57–76.

TARSKI Alfred (1932) 'Der Wahrheitsbegriff in den Sprachen der deduktiven Disziplinen', *Akademischer Anzeiger der Akademie der Wissenschaften in Wien, Mathematisch-naturwissenschaftliche Klasse* 69, pp. 23–25.

——— (1933a) *Pojęcie prawdy w jęykach nauk dedukcyjnych,* Varsovie. German edition, 'Der Wahrheitsbegriff in den formalisierten Sprachen', *Studia Philosophica* 1, 1936, pp. 261–405 (offprints dated 1935). English edition and translation by J. Woodger as 'The Concept of Truth in Formalized Languages', in Tarski (1956), pp. 152–278.

——— (1933b) 'Einige Betrachtungen über die Begriffe der ω-Wiederspruchsfreiheit und der ω-Vollständigkeit', *Monatshefte für Mathematik und Physik* 40, pp. 97–112. Translated by J. Woodger as 'Some Observations on the Concepts of ω-consistency and ω-completeness', in Tarski (1956), pp. 279–95.

——— (1936a) 'O ugruntowaniu naukowej semantyki', *Przegląd Filozoficzny* 39, pp. 50–7. Translated into German as 'Grundlegung der wissenschaftlichen Semantik', *Actes du Congrès International de Philosophie Scientifique, Paris 1935,* fasc. III, 'Langage et pseudo-problèmes', Paris: Hermann, 1936, pp. 1–8. Translated by J. Woodger as 'The Establishment of Scientific Semantics', in Tarski (1956), pp. 401–8.

——— (1936b) 'O pojęciu wynikana logicznego', *Przegląd Filozoficzny* 39, pp. 58–68. Translated into German as 'Über den Begriff der logischen Folgerung', *Actes du Congrès International de Philosophie Scientifique, Sorbonne, Paris 1935,* fasc. VII, 'Induction et Probabilité', Paris: Hermann, 1936, pp. 1–11. Translated by J. Woodger as 'On the Concept of Logical Consequence', in Tarski (1956), pp. 409–20.

——— (1956) *Logic, Semantics, Metamathematics,* ed. J. Woodger, Oxford: Oxford University Press, 2nd ed. with an introduction by J. Corcoran, Indiana: Hackett Publishing Company, 1983.

——— (1986) 'What are Logical Notions?', ed. J. Corcoran, *History and Philosophy of Logic* 7, pp. 143–54.

UEBEL Thomas (1992) 'Rational Reconstruction as Elucidation? Carnap in the Early Protocol Sentence Debate', *Synthese* 93, pp. 107–40.

——— (1995) 'Physicalism in Wittgenstein and the Vienna Circle', in K. Gavroglu, J. Stachel, and M. Wartofsky (eds), *Physics, Philosophy and the Scientific Community. Festschrift for Robert S. Cohen,* Dordrecht: Kluwer, 1995, pp. 328–56.

——— (2001) 'Carnap and Neurath in Exile: Can Their Disputes Be Resolved?', *International Studies in the Philosophy of Science* 15, pp. 211–20.

——— (2004) 'Carnap, the Left Vienna Circle, and Neopositivist Antimetaphysics', in Awodey and Klein (2004), pp. 247–77.

——— (2005) 'Learning Logical Tolerance: Hans Hahn on the Foundations of Mathematics', *History and Philosophy of Logic* 26, pp. 175–209.

——— (2007a) *Empiricism at the Crossroads: The Vienna Circle's Protocol Sentence Debate Revisited*, Chicago: Open Court.

——— (2007b) 'Carnap and the Vienna Circle: Rational Reconstructionism Refined', in Friedman and Creath (2007), pp. 153–75.

WAISMANN Friedrich, ed. (1967) *Wittgenstein und der Wiener Kreis*, Oxford: Blackwell.

WHITEHEAD Alfred, RUSSELL Bertrand (1910–12) *Principia Mathematica*, 3 vols, Cambridge: Cambridge University Press, 2nd ed. 1927.

WITTGENSTEIN Ludwig (1921) 'Logisch-Philosophische Abhandlung', *Annalen der Naturphilosophie* 14, pp. 185–262. Translated by C. K. Ogden as *Tractatus Logico-Philosophicus*, London: Kegan Paul, Trench, Trubner, 1922, and by D. Pears and B. McGuinness, London: Routledge, 1961.

——— (1922) *Tractatus Logico-Philosophicus*, London: Kegan Paul, Trench, Trubner.

——— (1980) *Wittgenstein's Lectures, Cambridge 1930–32*, from the notes of John King and Desmond Lee, ed. J. King, Totowa, NJ: Rowman & Littlefield.

# Index of Names

# Index of Subjects